NURSES'
ILLUST
PHYSIOLOGY

BY
Ann B. McNaught
M.B., Ch.B., Ph.D.
The University of Glasgow.

AND

Robin Callander
F.F.Ph., F.M.A.A., A.I.M.B.I.
Director, Medical Illustration Unit,
The University of Glasgow.

FOURTH EDITION

EDINBUR   RK 1983

AS
ALWAYS

To my husband, James A. Gilmour

Ann B. McNaught

To my wife, Elizabeth

Robin Callander

CHURCHILL LIVINGSTONE
Medical Division of Longman Group UK Limited

Distributed in the United States of America by Churchill Livingstone
Inc., 1560 Broadway, New York, 10036 and by associated
companies, branches and representatives throughout the world.

First Edition	1964
Reprinted	1965
Reprinted	1966
Reprinted	1968
Second edition	1971
Reprinted	1973
Third edition	1975
Reprinted	1977
Reprinted	1980
Fourth edition	1983
Reprinted	1986
Reprinted	1989
Reprinted	1990

ISBN 0-443-02703-X

Produced by Longman Group (FE) Ltd
Printed in Hong Kong

PREFACE

Henry Ford once said "Nothing is particularly hard if you divide it into small jobs". For the study of Physiology that is what we have tried to do here.

Each page deals with a "small" aspect of the subject and – as with our larger book "Illustrated Physiology" – is complete in itself. In general a lead-picture for each chapter indicates the relative position in the body of the system concerned. On subsequent pages the anatomy of major organs is presented in 3-dimensional form. Microscopic features are shown by inset with approximate magnifications indicated. Associated functions of the various parts are summarized alongside. Thereafter line diagrams are used, where applicable, to show how these functions are carried out. Footnotes indicate the method of regulation and control. Cross references have been inserted to pages where limitations of space make it difficult to incorporate those functions of an organ not directly involved in the work of the system under review.

The book is not intended to replace standard works on the subject much less to supplant lecture notes. Rather it is hoped that it may be thought suitable to augment the latter and perhaps save both tutor and student nurse the sometimes laborious task of copying blackboard diagrams. Its treatment has been found particularly useful for purposes of revision and review especially for those whose available study time tends to come in small rations.

We are indebted as before to a number of colleagues and sources acknowledged where possible in the text.

Glasgow, 1983

Ann B. NcNaught
Robin Callander

CONTENTS

4. Respiratory system (68–77)

5. Excretory system (78–86)

6. Endocrine system (87–100)

7. Reproductive system (101–115)

8. Nervous and locomotor systems (116–150)

CHAPTER 1.

Introduction:– <u>The TISSUES</u>

<u>The AMOEBA</u>

All living things are made of PROTOPLASM. Protoplasm exists in MICROSCOPIC UNITS called CELLS. The SIMPLEST living creatures consist of ONE CELL. The AMOEBA (which lives in pond water) exemplifies the BASIC STRUCTURE of all animal cells and shows the PHENOMENA which distinguish living from non-living things.

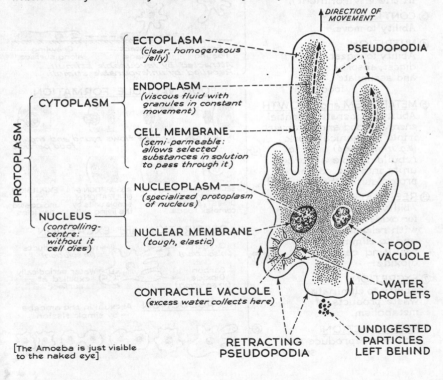

DIRECTION OF MOVEMENT

ECTOPLASM
(clear, homogeneous jelly)

PSEUDOPODIA

ENDOPLASM
(viscous fluid with granules in constant movement)

CYTOPLASM

CELL MEMBRANE
(semi-permeable: allows selected substances in solution to pass through it)

PROTOPLASM

NUCLEOPLASM
(specialized protoplasm of nucleus)

NUCLEAR MEMBRANE
(tough, elastic)

NUCLEUS
(controlling-centre: without it cell dies)

FOOD VACUOLE

CONTRACTILE VACUOLE
(excess water collects here)

WATER DROPLETS

[The Amoeba is just visible to the naked eye]

RETRACTING PSEUDOPODIA

UNDIGESTED PARTICLES LEFT BEHIND

The PHENOMENA which characterize all living things ----- are shown by the AMOEBA

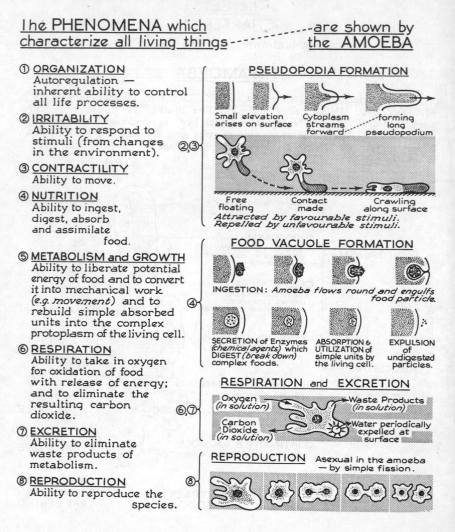

① **ORGANIZATION**
Autoregulation — inherent ability to control all life processes.

② **IRRITABILITY**
Ability to respond to stimuli (from changes in the environment).

③ **CONTRACTILITY**
Ability to move.

④ **NUTRITION**
Ability to ingest, digest, absorb and assimilate food.

⑤ **METABOLISM and GROWTH**
Ability to liberate potential energy of food and to convert it into mechanical work (*e.g. movement*) and to rebuild simple absorbed units into the complex protoplasm of the living cell.

⑥ **RESPIRATION**
Ability to take in oxygen for oxidation of food with release of energy; and to eliminate the resulting carbon dioxide.

⑦ **EXCRETION**
Ability to eliminate waste products of metabolism.

⑧ **REPRODUCTION**
Ability to reproduce the species.

PSEUDOPODIA FORMATION

Small elevation arises on surface

Cytoplasm streams forward

forming long pseudopodium

Free floating

Contact made

Crawling along surface

Attracted by favourable stimuli.
Repelled by unfavourable stimuli.

FOOD VACUOLE FORMATION

INGESTION: *Amoeba flows round and engulfs food particle.*

SECRETION of Enzymes (*chemical agents*) which DIGEST (*break down*) complex foods.

ABSORPTION & UTILIZATION of simple units by the living cell.

EXPULSION of undigested particles.

RESPIRATION and EXCRETION

Oxygen (*in solution*)

Carbon Dioxide (*in solution*)

Waste Products (*in solution*)

Water periodically expelled at surface

REPRODUCTION
Asexual in the amoeba — by simple fission.

2

The CELL

Higher animals, including MAN, are made up of millions of living cells which vary widely in structure and function but have certain features in common.

<u>A GENERALIZED ANIMAL CELL</u> *(of secretory type).* [The finer details are seen only in the high magnification afforded by ELECTRON MICROSCOPY]

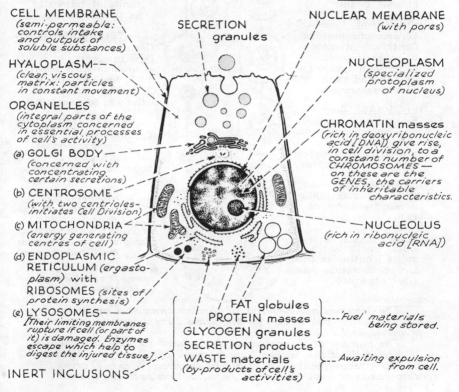

CYTOPLASM

CELL MEMBRANE
(semi-permeable: controls intake and output of soluble substances)

HYALOPLASM--
(clear, viscous matrix: particles in constant movement)

ORGANELLES
(integral parts of the cytoplasm concerned in essential processes of cell's activity)

(a) GOLGI BODY—
(concerned with concentrating certain secretions)

(b) CENTROSOME
(with two centrioles—initiates Cell Division)

(c) MITOCHONDRIA
(energy generating centres of cell)

(d) ENDOPLASMIC RETICULUM *(ergasto-plasm)* with RIBOSOMES *(sites of protein synthesis)*

(e) LYSOSOMES--
[Their limiting membranes rupture if cell (or part of it) is damaged. Enzymes escape which help to digest the injured tissue]

INERT INCLUSIONS

SECRETION
granules

NUCLEUS

NUCLEAR MEMBRANE
(with pores)

NUCLEOPLASM
(specialized protoplasm of nucleus)

CHROMATIN masses
(rich in deoxyribonucleic acid [DNA]) give rise, in cell division, to a constant number of CHROMOSOMES —on these are the GENES, the carriers of inheritable characteristics.

NUCLEOLUS
(rich in ribonucleic acid [RNA])

FAT globules
PROTEIN masses
GLYCOGEN granules } ---*'Fuel' materials being stored.*

SECRETION products
WASTE materials
(by-products of cell's activities) } -- *Awaiting expulsion from cell.*

CELL DIVISION (MITOSIS)

All cells arise from the division of pre-existing cells. In MITOSIS there is an exact QUALITATIVE division of the NUCLEUS and a less exact QUANTITATIVE division of the CYTOPLASM.

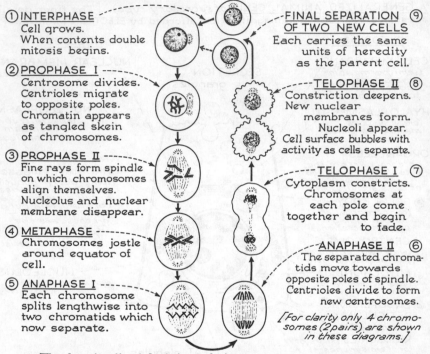

① INTERPHASE
Cell grows.
When contents double mitosis begins.

② PROPHASE I
Centrosome divides.
Centrioles migrate to opposite poles.
Chromatin appears as tangled skein of chromosomes.

③ PROPHASE II
Fine rays form spindle on which chromosomes align themselves.
Nucleolus and nuclear membrane disappear.

④ METAPHASE
Chromosomes jostle around equator of cell.

⑤ ANAPHASE I
Each chromosome splits lengthwise into two chromatids which now separate.

FINAL SEPARATION ⑨
OF TWO NEW CELLS
Each carries the same units of heredity as the parent cell.

TELOPHASE II ⑧
Constriction deepens.
New nuclear membranes form.
Nucleoli appear.
Cell surface bubbles with activity as cells separate.

TELOPHASE I ⑦
Cytoplasm constricts.
Chromosomes at each pole come together and begin to fade.

ANAPHASE II ⑥
The separated chromatids move towards opposite poles of spindle.
Centrioles divide to form new centrosomes.

[For clarity only 4 chromosomes (2 pairs) are shown in these diagrams.]

The longitudinal halving of chromosomes and genes ensures that each new cell receives the same hereditary factors as the original cell.

The number of chromosomes is constant for any one species.
The cells of the human body (somatic cells) carry 23 pairs — i.e. 46 chromosomes. A modified division (Meiosis) gives 23 chromosomes in ova and sperm. Fertilization restores the 46.

4

DIFFERENTIATION of ANIMAL CELLS

SPECIALIZATION distinguishes multicellular creatures from more primitive forms of life.

ONE-CELLED ANIMALS are capable of INDEPENDENT existence — they are *Undifferentiated* i.e. show all activities or ·····PHENOMENA of LIFE

MANY-CELLED ANIMALS	Cells CO-OPERATE for well-being of whole body.	*All cells retain powers of* ORGANIZATION
Differentiated ——	Groups of cells undergo adaptations and sacrifice some powers to fit them for special duties.	*IRRITABILITY* *NUTRITION* *METABOLISM* *RESPIRATION* *EXCRETION*

MODIFICATION of STRUCTURE ···· -*for efficient*- ···· SPECIALIZATION of FUNCTION ···· -*with*- ···· LOSS or REDUCTION of VERSATILITY
E.g.

SECRETORY CELL ········· Highly developed powers of SECRETION e.g. enzymes for chemical breakdown of foodstuffs.   *Diminished powers of* CONTRACTION and REPRODUCTION

Cytoplasm — Nucleus displaced to base by formed and stored secretion
Nucleus ··

FAT CELL ·············· STORAGE of FAT   *Loss of powers of* CONTRACTION and SECRETION

Cytoplasm — Cytoplasm displaced by stored fat
Nucleus

MUSCLE CELL ············ Highly developed powers of CONTRACTILITY   *Diminished powers of* SECRETION and REPRODUCTION

Cytoplasm — Nucleus
Elongated cell body

NERVE CELL ············ Highly developed powers of IRRITABILITY   *Loss of powers of* REPRODUCTION
Cytoplasm

Cytoplasm drawn out into long branching processes

(response to stimuli and transmission of impulses over long distances)   *i.e. if nerve cell is destroyed no regeneration is possible.*

Nucleus x400

ORGANIZATION OF TISSUES

Different cell types are not mixed haphazardly in the body.
Cells which are alike are arranged together to form TISSUES.
There are *four* main types of tissue:– 1. EPITHELIA or LINING,
2. CONNECTIVE or SUPPORTING, 3. MUSCULAR, 4. NERVOUS.

EPITHELIA

STRUCTURAL MODIFICATIONS *Sheets of cells with minimum intercellular substance*	SITE	SPECIALIZED FUNCTIONS *Line all internal and external surfaces of body*

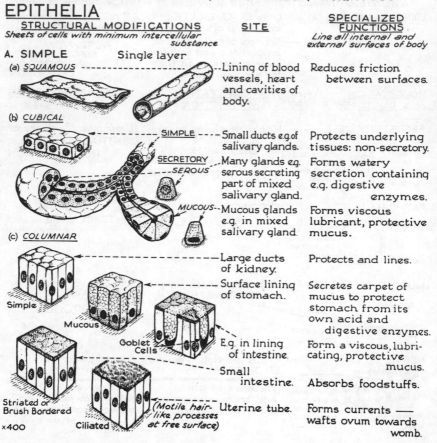

A. SIMPLE   Single layer

(a) *SQUAMOUS* — Lining of blood vessels, heart and cavities of body. — Reduces friction between surfaces.

(b) *CUBICAL*

SIMPLE — Small ducts e.g.of salivary glands. — Protects underlying tissues: non-secretory.

SECRETORY – SEROUS — Many glands e.g. serous secreting part of mixed salivary gland. — Forms watery secretion containing e.g. digestive enzymes.

MUCOUS — Mucous glands e.g. in mixed salivary gland. — Forms viscous lubricant, protective mucus.

(c) *COLUMNAR*

Simple — Large ducts of kidney. — Protects and lines.

Mucous — Surface lining of stomach. — Secretes carpet of mucus to protect stomach from its own acid and digestive enzymes.

Goblet Cells — E.g. in lining of intestine. — Form a viscous, lubricating, protective mucus.

— Small intestine. — Absorbs foodstuffs.

Striated or Brush Bordered

×400   Ciliated   (Motile hair-like processes at free surface) — Uterine tube. — Forms currents — wafts ovum towards womb.

6

B. PSEUDOSTRATIFIED COLUMNAR
CILIATED (with Goblet cells) ------- Respiratory passages.

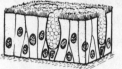

(a) Protective.

(b) Cilia form currents to move mucus-trapped particles towards the back of the throat.

C. COMPOUND - Many layers of cells
TRUE STRATIFIED
(a) TRANSITIONAL --------------------- Urinary passages.

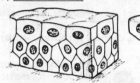

(a) Extensible.

(b) Protective — prevents penetration of urine into underlying tissues.

(b) STRATIFIED SQUAMOUS -------------- Surfaces subjected to great wear and tear.

Most resistant type of epithelium.

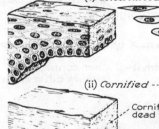

(i) Uncornified

Internal surfaces e.g. mouth and gullet.

Protects against friction.

(ii) Cornified ----------- External skin surfaces.

Protects against wear and tear, evaporation and extremes of temperature.

- Cornified dead cell layers
- Clear layer
- Granular layer
- Prickle cell layers
- Germinal layer (for cell replacement)

×400

[Dead cell layers vary in thickness — e.g. thickest on soles of feet and palms of hands.]

7

CONNECTIVE TISSUES

STRUCTURAL MODIFICATIONS	SPECIALIZED FUNCTIONS
Cells plus large amount of intercellular matrix and extracellular elements.	*Form framework, connecting, supporting and packing tissues of the body.*

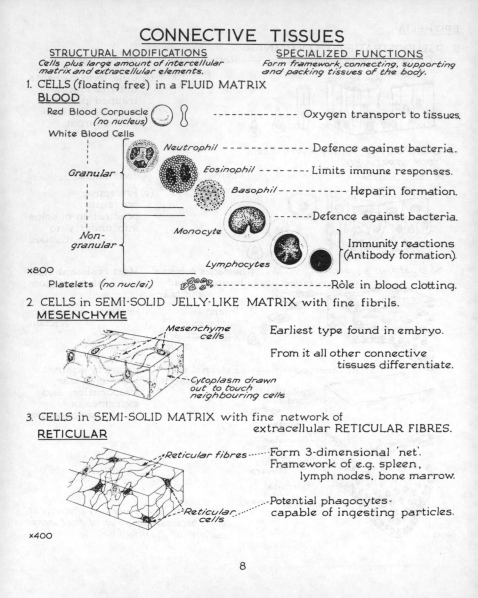

1. CELLS (floating free) in a FLUID MATRIX
   BLOOD

   Red Blood Corpuscle *(no nucleus)* ------------ Oxygen transport to tissues.

   White Blood Cells

   Granular
   - Neutrophil ------------ Defence against bacteria.
   - Eosinophil -------- Limits immune responses.
   - Basophil ----------- Heparin formation.

   Non-granular
   - Monocyte ------- Defence against bacteria.
   - Lymphocytes } Immunity reactions (Antibody formation).

   x800

   Platelets *(no nuclei)* ------------------ Rôle in blood clotting.

2. CELLS in SEMI-SOLID JELLY-LIKE MATRIX with fine fibrils.
   MESENCHYME

   Mesenchyme cells

   Earliest type found in embryo.

   From it all other connective tissues differentiate.

   Cytoplasm drawn out to touch neighbouring cells

3. CELLS in SEMI-SOLID MATRIX with fine network of extracellular RETICULAR FIBRES.
   RETICULAR

   Reticular fibres ...... Form 3-dimensional 'net'. Framework of e.g. spleen, lymph nodes, bone marrow.

   Reticular cells ...... Potential phagocytes - capable of ingesting particles.

   x400

8

CONNECTIVE TISSUES

4. CELLS in SEMI-SOLID MATRIX with thicker collagenous
and elastic fibres.

(a) LOOSE FIBROUS

Fibroblasts (cells actively forming fibres)
Collagen fibres
Elastic fibres

"Packing" tissue between organs; sheaths of muscles, nerves and blood vessels.

(b) DENSE FIBROUS

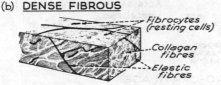

Fibrocytes (resting cells)
Collagen fibres
Elastic fibres

Strong, inelastic yet pliable— e.g. ligaments; capsules of joints; heart valves.

(c) ELASTIC

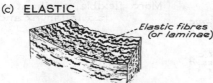

Elastic fibres (or laminae)

Strong, extensible and flexible— e.g. in walls of blood vessels and air passages.

(d) ADIPOSE

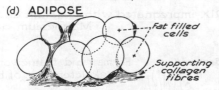

Fat filled cells
Supporting collagen fibres

Protective "cushion" for organs. Insulating layer in skin. Storage of fat reserves.

(e) TENDON

Tendon cells
Dense fibrous tissue bundles

Tough, inelastic cords of dense fibrous tissue which attach muscles to bones.

x400

CONNECTIVE TISSUES

5. CELLS in SOLID ELASTIC MATRIX with fibres.

(a) HYALINE CARTILAGE

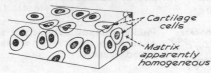

Cartilage cells

Matrix apparently homogeneous

Firm yet resilient
e.g. in air passages; ends of bones at joints.
Transition stage in bone
formation.

(b) WHITE FIBRO-CARTILAGE

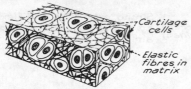

Cartilage cells

Collagen fibres in matrix

Tough. Resistant to stretching.
Acts as shock absorber between vertebrae.

(c) ELASTIC or YELLOW FIBRO-CARTILAGE

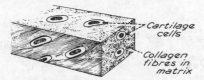

Cartilage cells

Elastic fibres in matrix

More flexible, resilient —
e.g. in larynx and ear.

×400

6. CELLS in SOLID RIGID MATRIX impregnated with Calcium and
Magnesium Salts.

BONE

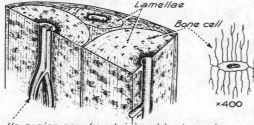

Lamellae

Bone cell

Forms rigid framework
(or skeleton)   of body.

×400

Haversian canal containing blood vessels.
× 75

10

MUSCULAR TISSUES

All have ELONGATED cells with special development of CONTRACTILITY.

1. SMOOTH, UNSTRIPED, VISCERAL or INVOLUNTARY muscle

Nucleus

Least specialized. Shows slow rhythmical contraction and relaxation.
Not under voluntary control.
Found in walls of viscera and blood vessels.

2. CARDIAC or HEART muscle

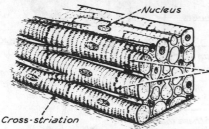

Nucleus

Cross-striation

More highly specialized.
Rapid rhythmical contraction (and relaxation) spreads through whole muscle mass.
Not under voluntary control.
Found only in heart wall.

→ *Cells adhere, end to end, at intercalated discs to form long 'fibres' which branch and connect with adjacent 'fibres'.*

(Note: There is no protoplasmic continuity between cells.)

3. STRIATED, STRIPED, SKELETAL or VOLUNTARY muscle

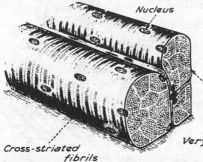

Nucleus

Cross-striated fibrils

×400

Most highly specialized.
Very rapid, powerful contractions of individual fibres.
Under voluntary control.
Found in e.g. muscles of trunk, limbs, head.

Thick covering membrane (sarcolemma)

Very long, large multi-nucleated units. No branching.

11

NERVOUS TISSUES

Nervous tissue is divided into:-
(a) **NEURONES** or *True* nerve cells specialized in *IRRITABILITY, CONDUCTION, INTEGRATION*

MOTOR – pass messages from brain and spinal cord to muscles and glands.
ASSOCIATION – relay messages between neurones.
SENSORY – receive and pass messages from environment to brain and spinal cord.

(b) *Supporting* cells e.g. NEUROGLIA in brain and spinal cord; SHEATH (*Schwann*) cells on nerve fibres outside brain and spinal cord.

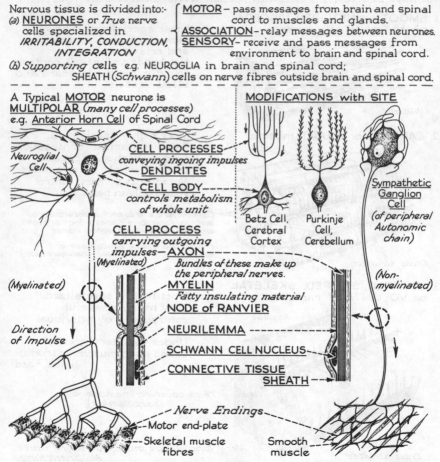

A Typical **MOTOR** neurone is **MULTIPOLAR** (*many cell processes*) e.g. <u>Anterior Horn Cell</u> of Spinal Cord

Neuroglial Cell

CELL PROCESSES
conveying ingoing impulses
— DENDRITES

CELL BODY
controls metabolism of whole unit

CELL PROCESS
carrying outgoing impulses — **AXON**
(Myelinated)

(Myelinated)

Direction of Impulse

Bundles of these make up the peripheral nerves.
MYELIN
Fatty insulating material
NODE of RANVIER

NEURILEMMA

SCHWANN CELL NUCLEUS

CONNECTIVE TISSUE SHEATH

Nerve Endings
-- Motor end-plate
-- Skeletal muscle fibres

MODIFICATIONS with SITE

Betz Cell, Cerebral Cortex

Purkinje Cell, Cerebellum

Sympathetic Ganglion Cell
(*of peripheral Autonomic chain*)

(Non-myelinated)

Smooth muscle

Most multipolar neurones are MOTOR (efferent) or ASSOCIATION in function.

NERVOUS TISSUES

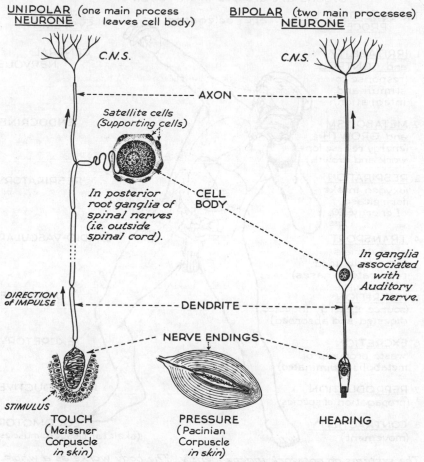

UNIPOLAR *(one main process
NEURONE* *leaves cell body)*

BIPOLAR *(two main processes)*
NEURONE

C.N.S.

C.N.S.

← ---------------------- AXON ---------------- →

*Satellite cells
(Supporting cells)*

CELL
BODY

*In posterior
root ganglia of
spinal nerves
(i.e. outside
spinal cord).*

*In ganglia
associated
with
Auditory
nerve.*

DIRECTION
of IMPULSE ↑

← ------ DENDRITE -------- →

← ----- NERVE ENDINGS ---- →

STIMULUS

TOUCH
*(Meissner
Corpuscle
in skin)*

PRESSURE
*(Pacinian
Corpuscle
in skin)*

HEARING

All unipolar and bipolar neurones are SENSORY (afferent) in function.

2

The BODY SYSTEMS

The tissues are arranged to form ORGANS.
Organs are grouped into SYSTEMS.

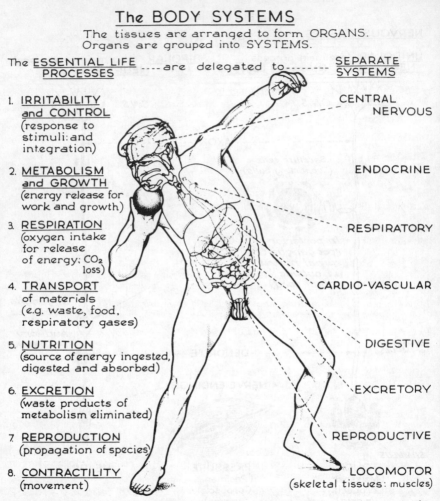

The <u>ESSENTIAL LIFE PROCESSES</u> ---- are delegated to -------- <u>SEPARATE SYSTEMS</u>

1. <u>IRRITABILITY and CONTROL</u> (response to stimuli; and integration)

2. <u>METABOLISM and GROWTH</u> (energy release for work and growth)

3. <u>RESPIRATION</u> (oxygen intake for release of energy; CO_2 loss)

4. <u>TRANSPORT</u> of materials (e.g. waste, food, respiratory gases)

5. <u>NUTRITION</u> (source of energy ingested, digested and absorbed)

6. <u>EXCRETION</u> (waste products of metabolism eliminated)

7. <u>REPRODUCTION</u> (propagation of species)

8. <u>CONTRACTILITY</u> (movement)

CENTRAL NERVOUS

ENDOCRINE

RESPIRATORY

CARDIO-VASCULAR

DIGESTIVE

EXCRETORY

REPRODUCTIVE

LOCOMOTOR (skeletal tissues; muscles)

The systems do not work independently. The body works as a whole. Health and well-being depend on the co-ordinated effort of every part.

14

CHAPTER 2

NUTRITION, METABOLISM AND DIGESTION

BASIC CONSTITUENTS of PROTOPLASM

Protoplasm is made up of certain <u>Elements</u> present mainly in <u>Chemical Combination</u>
Eg. in a 70kg man these four elements make up most of the body weight:-

HYDROGEN(H) 6,580 g	A large amount of the H and O is present as WATER (H_2O)	C,H and O combine chemically to form CARBOHYDRATES and FATS (*chief sources of ENERGY in living protoplasm*)	C,H,O and N combine to form PROTEINS (*the main BUILDING constituents of all protoplasm*)
OXYGEN (O) 43,550 g			
CARBON (C) 12,590 g			
NITROGEN (N) 1,815 g			

The following eight make up much of the remaining body weight:-

CALCIUM (Ca)   1,700g   } Important constituents of blood and of hard tissues-
PHOSPHORUS(P) 680g   }   e.g. bones and teeth.
CHLORINE (Cl)   115g   } Important constituents of body fluids.
SODIUM (Na)   70g   }
POTASSIUM (K)   70g   Important constituent of all cells.
SULPHUR (S)   100g
MAGNESIUM (Mg) 42g   Important for activity of brain, nerves and muscles.
IRON (Fe)   7g   Important component of haemoglobin in red blood cells.

Trace elements, such as manganese, copper, iodine, zinc, cobalt, fluorine, strontium, *make up the last few grams or so.*

Note:- *Apart from WATER, the chief constituents in both plant and animal protoplasm are present as compounds of CARBON, i.e. they are ORGANIC SUBSTANCES.*

<u>CARBOHYDRATES</u> — the simpler ones are SUGARS ; the more complex- STARCH (made up of hundreds of units of sugar tied chemically together). They consist of atoms of C, H and O.

<u>FATS</u> — the molecule is made up of smaller molecules of FATTY ACIDS linked chemically with a molecule of GLYCEROL. They consist of atoms of the elements C, H and O.

<u>PROTEINS</u> — chief organic constituents of all protoplasm — the molecule is made up of smaller units called AMINO ACIDS. These consist of atoms of C,H,O,N and sometimes S and P.

There are thousands of different kinds of Protein but only about 20 different amino acids. Differences between proteins depend on the amino acids present and on their number and arrangement.

SOURCE of ENERGY: PHOTOSYNTHESIS

The essential life processes or the phenomena which characterize life depend on the use of ENERGY. The SUN is the SOURCE of ENERGY for ALL LIVING THINGS.

Only GREEN PLANTS can TRAP and STORE the sun's energy and build from relatively simple substances the energy-rich and body-building compounds — *CARBOHYDRATES, FATS and PROTEINS* — required by PROTOPLASM. This process is called <u>PHOTOSYNTHESIS</u>.

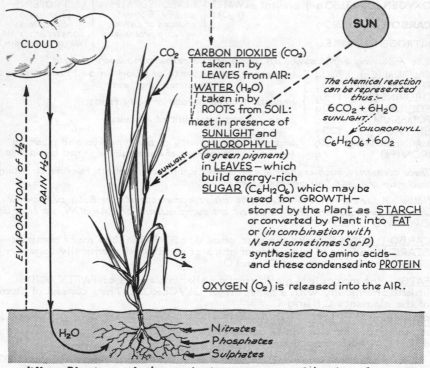

CLOUD

CO₂

EVAPORATION of H₂O

RAIN H₂O

SUNLIGHT

O₂

H₂O

<u>CARBON DIOXIDE</u> (CO_2) taken in by LEAVES from AIR:
<u>WATER</u> (H_2O) taken in by ROOTS from SOIL:
meet in presence of <u>SUNLIGHT</u> and <u>CHLOROPHYLL</u> *(a green pigment)* in <u>LEAVES</u> — which build energy-rich <u>SUGAR</u> ($C_6H_{12}O_6$) which may be used for GROWTH — stored by the Plant as <u>STARCH</u> or converted by Plant into <u>FAT</u> or *(in combination with N and sometimes S or P)* synthesized to amino acids — and these condensed into <u>PROTEIN</u>

<u>OXYGEN</u> (O_2) is released into the AIR.

SUN

The chemical reaction can be represented thus:-

$$6CO_2 + 6H_2O$$
SUNLIGHT →
CHLOROPHYLL
$$C_6H_{12}O_6 + 6O_2$$

Nitrates
Phosphates
Sulphates

When Plants or their products are eaten this stored energy becomes available to animals and man.

16

CARBON 'CYCLE' in NATURE

Animal bodies are unable to build *Proteins, Carbohydrates* or *Fats* from the raw materials. They must be built up for them by *Plants*.

CARBON is the basic building unit of all these compounds.

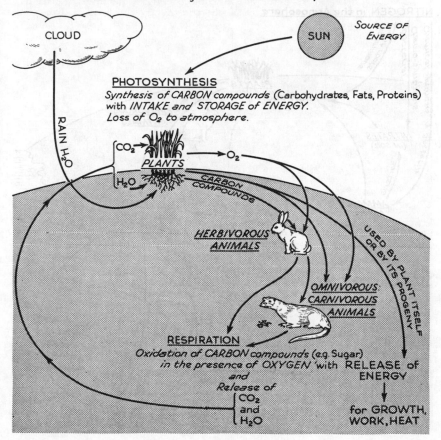

CLOUD

SUN

SOURCE OF ENERGY

PHOTOSYNTHESIS
Synthesis of CARBON compounds (Carbohydrates, Fats, Proteins) *with INTAKE and STORAGE of ENERGY.*
Loss of O_2 to atmosphere.

RAIN H_2O

CO_2

PLANTS

H_2O

O_2

CARBON COMPOUNDS

HERBIVOROUS ANIMALS

USED BY PLANT ITSELF OR BY ITS PROGENY

OMNIVOROUS: CARNIVOROUS ANIMALS

RESPIRATION
Oxidation of CARBON compounds (e.g. Sugar)
in the presence of OXYGEN 'with RELEASE of ENERGY'
and
Release of
CO_2
and
H_2O

for GROWTH, WORK, HEAT

NITROGEN 'CYCLE' in NATURE

Although animals are surrounded by NITROGEN in the air they can only use the NITROGEN trapped by Plants.

<u>NITROGEN in the Atmosphere</u>

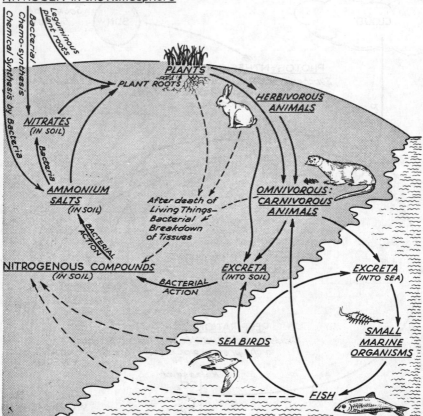

NUTRITION

Man eats as <u>FOOD</u> the substances which PLANTS (and, through them, ANIMALS) have made. These are broken down in man's body into simpler chemical units which provide

BUILDING and PROTECTIVE MATERIALS

Man requires more or less the same elements as plants. *(Some, such as iodine, sodium, iron, calcium, he assimilates in INORGANIC FORM.)*

Most must be built up for him by plants:-

ORGANICALLY COMBINED
Carbon, Nitrogen, and Sulphur, etc.
ESSENTIAL AMINO ACIDS

ESSENTIAL FATTY ACIDS
Certain VITAMINS

These are
<u>USED to</u>
<u>BUILD, MAINTAIN or REPAIR</u>
PROTOPLASM

- -

BODY BUILDING REQUIREMENTS
of the INDIVIDUAL determine

<u>QUALITY</u>
of DIET

ENERGY

Stored originally by plants <u>RELEASED</u> in man's cells by <u>OXIDATION</u>

When food is 'burned' it gives up its stored energy

PROTEINS	yield 4·1 kilocalories	*Units of*
CARBOHYDRATES	yield 4·1 kilocalories	*Energy*
FATS	yield 9·2 kilocalories	*per gram*

Most of this appears as <u>HEAT</u>
and is <u>USED for</u>
<u>KEEPING BODY WARM</u>;
some is used for <u>WORK</u> *of Cells*

- -

ENERGY REQUIREMENTS
of the INDIVIDUAL determine

<u>QUANTITY</u>
of DIET

For a
BALANCED DIET
TOTAL INTAKE
of
Essential Constituents and Energy Units
must balance...
...Amounts *stored* plus amounts *lost* from body plus amounts *used*
as Work or Heat.

ENERGY-GIVING FOODS

All the main foodstuffs yield ENERGY - the energy originally trapped by Plants. CARBOHYDRATES and FATS are the chief energy-giving foods. PROTEINS can give energy but are mainly used for building and repairing protoplasm.

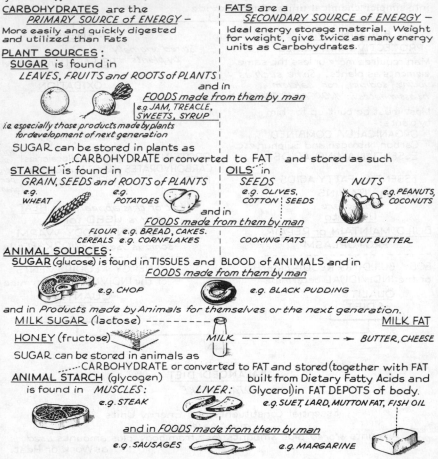

CARBOHYDRATES are the
PRIMARY SOURCE of ENERGY –
More easily and quickly digested
and utilized than Fats

FATS are a
SECONDARY SOURCE of ENERGY –
Ideal energy storage material. Weight
for weight, give twice as many energy
units as Carbohydrates.

PLANT SOURCES:

SUGAR is found in
LEAVES, FRUITS and ROOTS of PLANTS

and in
FOODS made from them by man
e.g. JAM, TREACLE,
SWEETS, SYRUP

i.e. especially those products made by plants
for development of next generation

SUGAR can be stored in plants as
CARBOHYDRATE or converted to FAT and stored as such

STARCH is found in
GRAIN, SEEDS and ROOTS of PLANTS
e.g.
WHEAT   e.g.
POTATOES

and in
FOODS made from them by man
FLOUR e.g. BREAD, CAKES.
CEREALS e.g. CORNFLAKES

OILS in
SEEDS   NUTS
e.g. OLIVES,   e.g. PEANUTS,
COTTON SEEDS   COCONUTS

COOKING FATS   PEANUT BUTTER

ANIMAL SOURCES:

SUGAR (glucose) is found in TISSUES and BLOOD of ANIMALS and in
FOODS made from them by man
e.g. CHOP   e.g. BLACK PUDDING

and in *Products made by Animals for themselves or the next generation.*

MILK SUGAR (lactose) ---------   MILK FAT

HONEY (fructose)   MILK ------------→ BUTTER, CHEESE

SUGAR can be stored in animals as
CARBOHYDRATE or converted to FAT and stored (together with FAT

ANIMAL STARCH (glycogen)   built from Dietary Fatty Acids and
is found in *MUSCLES:*   *LIVER:*   Glycerol) in FAT DEPOTS of body.
e.g. STEAK   e.g. SUET, LARD, MUTTON FAT, FISH OIL

and in *FOODS made from them by man*
e.g. SAUSAGES   e.g. MARGARINE

BODY-BUILDING FOODS

PROTEIN is the chief body-building food. Because it is the chief constituent of protoplasm., tissues of Plants and Animals are the richest sources.

PLANT SOURCES

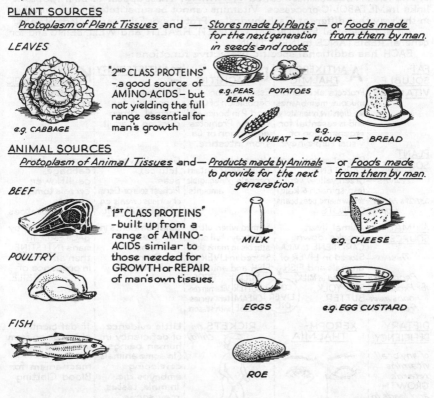

Protoplasm of Plant Tissues and — *Stores made by Plants — or Foods made for the next generation from them by man.* in *seeds and roots*

LEAVES

"2ND CLASS PROTEINS" –a good source of AMINO-ACIDS– but not yielding the full range essential for man's growth

e.g. CABBAGE

e.g. PEAS, BEANS

POTATOES

WHEAT → e.g. FLOUR → BREAD

ANIMAL SOURCES

Protoplasm of Animal Tissues and— *Products made by Animals — or Foods made to provide for the next generation from them by man.*

BEEF

"1ST CLASS PROTEINS" –built up from a range of AMINO-ACIDS similar to those needed for GROWTH or REPAIR of man's own tissues.

POULTRY

MILK

e.g. CHEESE

EGGS

e.g. EGG CUSTARD

FISH

ROE

These foods between them also contain other important body-building elements:- e.g. CALCIUM and PHOSPHORUS for making BONES and TEETH hard. IRON for building HAEMOGLOBIN– the OXYGEN CARRYING substance in RED BLOOD CELLS. IODINE for building the THYROID HORMONE.

PROTECTIVE FOODS — I

Even when provided with BODY-BUILDING and ENERGY-GIVING FOODS, human tissues cannot build or repair their protoplasm or release the energy in the absence of 'PROTECTIVE' substances called VITAMINS. These appear to be important links in METABOLIC processes. Vitamins cannot be substituted for one another in the way that different Carbohydrates, Proteins or Fats can replace others.

All are essential for NORMAL GROWTH, HEALTH and WELL-BEING and for general RESISTANCE to INFECTION.

EACH has additional specific protective functions:-

FAT-SOLUBLE VITAMINS	A ANTIXEROPH-THALMIC	D ANTI-RACHITIC	E ANTISTERILITY	K ANTI-HAEMORRHAGIC
	Protects skin and mucous membranes (e.g. digestive, respiratory) and is essential for regeneration of Visual Purple in eye.	Essential for deposition of Ca and P in bones and teeth. Promotes absorption of Ca from intestine.	Perhaps important for normal repro-duction in adult.	Important for normal blood clotting.
PLANT SOURCES	Present as precursor CAROTENE.	Vegetables, fruits and cereals contain negligible amounts	Green leaves (e.g. lettuce), peas. Richest source-Germ of various cereals e.g. WHEAT GERM OIL.	Spinach, kale, cabbage, cauliflower, cereals, tomatoes, carrots, potatoes.
LEAVES—	Green, yellow veg. (esp. spinach & kale)			
SEEDS & FRUITS—	Yellow maize, peas, beans			
Roots—	CARROTS			
ANIMAL SOURCES	Animal liver breaks down CAROTENE to Vit.A Stored in LIVER of animals and FISH	Formed when ultra-violet rays fall on sterols in man's skin. Stored in LIVER of FISH and animals	Small amounts in Meat and Dairy Produce	Synthesized by BACTERIA in man's INTESTINE, then absorbed in presence of BILE.
Tissues—				
PRODUCTS— for Progeny Foods made from them	Secreted in MILK, EGG YOLK. BUTTER. CREAM.	COD LIVER OIL	MILK BUTTER CREAM CHEESE	EGGS Vitamin con-tent varies with season.
DIETARY DEFICIENCY	XEROPH-THALMIA	RICKETS (in child)	Little evidence of deficiency in human beings. (In some animals, developing embryos die. In male, testes may show degenerative changes.)	If deficient absorption from intestine → upset in mechanism for Blood Clotting
of any or all vitamins retards GROWTH and leads to SEVERE DISEASE and eventually DEATH		OSTEOMALACIA (in adult)		

22

PROTECTIVE FOODS — II

WATER-SOLUBLE VITAMINS

B Complex ---

Together form essential parts of Enzymes concerned in metabolism and release of energy by cells

Includes
B₁ (*Aneurin, Thiamine*)
ANTINEURITIC
B₂ RIBOFLAVIN
NICOTINIC ACID
} concerned with Release of Energy from Foodstuffs.

B₆ PYRIDOXINE – concerned in metabolism of certain amino acids.

PANTOTHENIC ACID

BIOTIN – concerned in carbohydrate metabolism.

CHOLINE] probably promote
INOSITOL ∫ utilization of fats.

PARA-AMINOBENZOIC ACID – no known rôle in man.

FOLIC ACID – an ANTIANAEMIA factor.

B₁₂ COBALAMINE – ANTIPERNICIOUS ANAEMIA factor – involved in formation of red blood cells.

C ANTISCORBUTIC

Essential for formation and maintenance of intercellular cement and connective tissue.

PLANT SOURCES

Common natural sources for most members of B Complex:— Pulses e.g. green peas (B₁,B₂), seeds and OUTER COATS OF GRAIN, e.g. rice, wheat (B₁,B₂,Nic.A). Nuts, e.g. peanuts (B₁). Yeast and yeast extracts (B₁,B₂,Nicotinic Acid).

Green veg. e.g. parsley. Citrus fruits, e.g. oranges, lemons. Tomatoes, ROSEHIPS, blackcurrants. Peas. Potatoes.

ANIMAL SOURCES

B₂ and Nicotinic Acid are synthesized by BACTERIA in the human intestine. Found in LIVER (B₁,B₂,B₁₂) Nicotinic Acid,etc), bacon and lean meat, meat extract, milk (B₂, Nic.Acid), eggs (B₂), cheese (B₂).

Stored in body – High concentration in Adrenal Glands. Found in meat, liver. Secreted in milk.

DIETARY DEFICIENCY

Members of B Complex frequently absent together giving grave disturbances of chemical reactions in all tissues.

B₁ BERI-BERI
Disorder of Nervous System.
{ Muscular Paralysis and Weakness.
disturbance in Sensation.
Oedema.
Gastro-intestinal upsets.
Heart Failure.

B₁₂ PERNICIOUS ANAEMIA

B₂ RIBOFLAVINOSIS
Roughening of skin. Cornea becomes cloudy. Cracks and fissures around lips and tongue.

NICOTINIC ACID PELLAGRA
Roughening and reddening of skin. Tongue red and sore. Gastro-intestinal upsets. Mental derangement.

SCURVY
Intercellular cement breaks down.
Capillary walls leak.
Gums swell and bleed easily.
Wounds heal slowly.

ENERGY REQUIREMENTS *MALE*

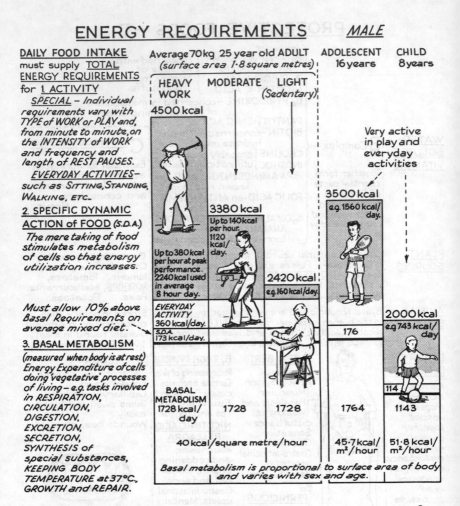

DAILY FOOD INTAKE must supply <u>TOTAL ENERGY REQUIREMENTS</u> for <u>1. ACTIVITY</u>

   SPECIAL – *Individual requirements vary with TYPE of WORK or PLAY and, from minute to minute, on the INTENSITY of WORK and frequency and length of REST PAUSES.*

   EVERYDAY ACTIVITIES – *such as SITTING, STANDING, WALKING, ETC.*

2. SPECIFIC DYNAMIC ACTION of FOOD (S.D.A.)

The mere taking of food stimulates metabolism of cells so that energy utilization increases.

Must allow 10% above Basal Requirements on average mixed diet. --->

3. BASAL METABOLISM

(measured when body is at rest) *Energy Expenditure of cells doing 'vegetative' processes of living – e.g. tasks involved in* RESPIRATION, CIRCULATION, DIGESTION, EXCRETION, SECRETION, SYNTHESIS *of special substances,* KEEPING BODY TEMPERATURE at 37°C., GROWTH and REPAIR.

Average 70 kg 25 year old ADULT *(surface area 1·8 square metres)*   ADOLESCENT 16 years   CHILD 8 years

HEAVY WORK	MODERATE	LIGHT (Sedentary)		
4500 kcal				
	3380 kcal		3500 kcal e.g. 1560 kcal/day.	
Up to 380 kcal per hour at peak performance. 2240 kcal used in average 8 hour day.	Up to 140 kcal per hour. 1120 kcal/day.	2420 kcal e.g. 160 kcal/day.		2000 kcal e.g. 743 kcal/day
	EVERYDAY ACTIVITY 360 kcal/day. S.D.A. 173 kcal/day.		176	114
BASAL METABOLISM 1728 kcal/day	1728	1728	1764	1143
40 kcal/square metre/hour			45·7 kcal/ m²/hour	51·8 kcal/ m²/hour

Very active in play and everyday activities

Basal metabolism is proportional to surface area of body and varies with sex and age.

[All figures are approximate and intended only as a general guide.]

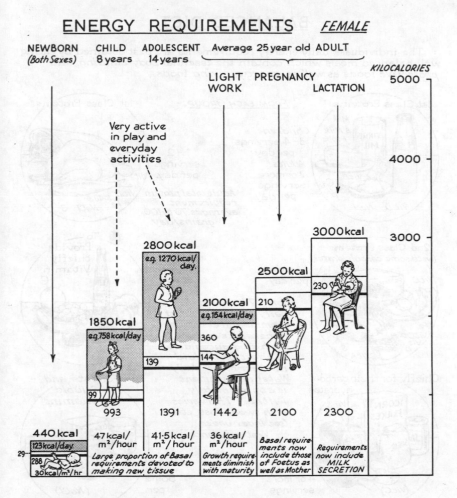

ENERGY REQUIREMENTS *FEMALE*

NEWBORN *(Both Sexes)* CHILD 8 years ADOLESCENT 14 years Average 25 year old ADULT

KILOCALORIES

LIGHT WORK PREGNANCY LACTATION

5000

Very active in play and everyday activities

4000

3000kcal

3000

2800 kcal
e.g. 1270 kcal/day.

2500kcal
230

2100kcal
e.g.154 kcal/day
360

210

1850 kcal
e.g.758 kcal/day

144

139

99

993 1391 1442 2100 2300

440 kcal
123kcal/day.
29
288
30kcal/m²/hr

47 kcal/m²/hour
Large proportion of Basal requirements devoted to making new tissue

41.5 kcal/m²/hour

36 kcal/m²/hour
Growth requirements diminish with maturity

Basal requirements now include those of Foetus as well as Mother

Requirements now include MILK SECRETION

[Proportion needed for growth diminishes in both sexes with age.]

25

BALANCED DIET

The individual's daily energy requirements are best obtained by eating well-balanced meals which contain the essential *body-building* and *protective* foods as well as *energy-giving* foods.

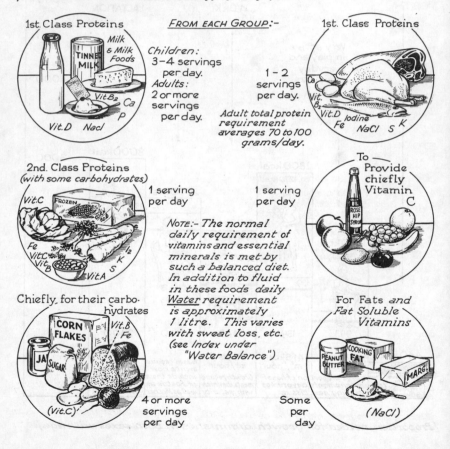

1st. Class Proteins

Milk & Milk Foods

TINNED MILK

Vit.B₂ Ca P

Vit.D Nacl

FROM EACH GROUP:-

Children:
3–4 servings per day.
Adults:
2 or more servings per day.

1 – 2 servings per day.

Adult total protein requirement averages 70 to 100 grams/day.

1st. Class Proteins

Ca Vit. B₂

Vit.D Fe Iodine NaCl S K

2nd. Class Proteins
(with some carbohydrates)

Vit.C FROZEN

Fe Vit.C Vit.B Vit.A

1 serving per day

1 serving per day

NOTE:- The normal daily requirement of vitamins and essential minerals is met by such a balanced diet. In addition to fluid in these foods daily <u>Water</u> *requirement is approximately 1 litre. This varies with sweat loss, etc. (see Index under "Water Balance")*

To Provide chiefly Vitamin C

ROSE HIP SYRUP

Chiefly, for their carbo-hydrates

CORN FLAKES

JA SUGAR

Vit.B Fe

(Vit.C)

4 or more servings per day

Some per day

For Fats *and* Fat Soluble *Vitamins*

PEANUT BUTTER COOKING FAT MARG.

(NaCl)

DIGESTIVE SYSTEM

The ALIMENTARY CANAL and ASSOCIATED GLANDS } Special System for dealing with <u>FOOD and FLUIDS</u>

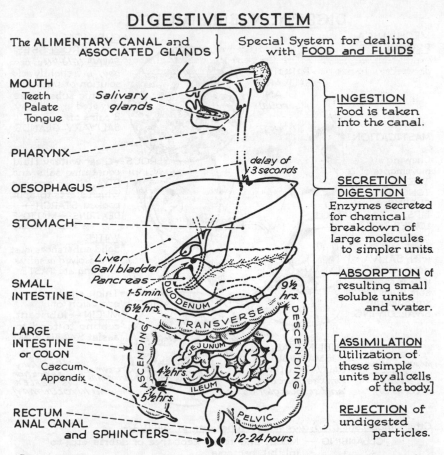

MOUTH
Teeth
Palate
Tongue

Salivary glands

PHARYNX

delay of 3 seconds

OESOPHAGUS

STOMACH

Liver
Gall bladder
Pancreas
1-5 min.

6½ hrs.

DUODENUM

9½ hrs.

SMALL INTESTINE

TRANSVERSE

ASCENDING

DESCENDING

LARGE INTESTINE or COLON
Caecum
Appendix

JEJUNUM

4½ hrs.

ILEUM

5½ hrs.

RECTUM
ANAL CANAL
and SPHINCTERS

PELVIC

12-24 hours

<u>INGESTION</u>
Food is taken into the canal.

<u>SECRETION</u> & <u>DIGESTION</u>
Enzymes secreted for chemical breakdown of large molecules to simpler units.

<u>ABSORPTION</u> of resulting small soluble units and water.

<u>ASSIMILATION</u>
Utilization of these simple units by all cells of the body.]

<u>REJECTION</u> of undigested particles.

[*Times taken by first part of meal to reach various parts of canal after leaving mouth are indicated. Latter part takes 3-5 hrs. to leave stomach.*]

During its progress along the canal FOOD is subjected to MECHANICAL as well as CHEMICAL changes to make it suitable for absorption into blood stream.

27

DIGESTION in the MOUTH

MECHANICAL PROCESSES

MASTICATION

Chewing movements of TEETH, TONGUE, CHEEKS, LIPS, LOWER JAW, break down food, mix it with SALIVA and roll it into a moist soft mass *(BOLUS)* suitable for SWALLOWING

[Mastication is under VOLUNTARY control]

PAROTID SALIVARY GLAND *(serous)*

×200

SUBLINGUAL, and SUBMANDIBULAR SALIVARY GLANDS *(mixed mucous and serous glands)*

Crescent of Serous cells

×200

CHEMICAL PROCESSES

SALIVA *(1-1½ litres per day)* is a slightly acid solution of salts and organic substances secreted mainly by 3 pairs of SALIVARY GLANDS.

SEROUS ACINI → Clear watery fluid containing salts and PTYALIN – an enzyme which starts to split cooked STARCH → DEXTRINS → MALTOSE *(a simpler sugar)*

WATER
Solid substances must be dissolved in saliva to stimulate TASTE buds.

MUCOUS ACINI → Thick slimy secretion of MUCIN → lubricant coating to food to assist SWALLOWING

[Salivary secretion is REFLEX and INVOLUNTARY]

Other important *(non-digestive)* functions of SALIVA :-

CLEANSING — Mouth and teeth kept free of debris, etc., to inhibit bacteria.

MOISTENING and LUBRICATING — Soft parts of mouth kept pliable for SPEECH.

EXCRETORY— Many organic substances *(e.g. urea, sugar)* and inorganic substances *(e.g. mercury, lead)* can be excreted in saliva.

CONTROL of SALIVARY SECRETION

Increased secretion at mealtimes is REFLEX *(involuntary)*.

Salivary Reflexes are of two types:–

(a) <u>UNCONDITIONED</u> *(inborn)*

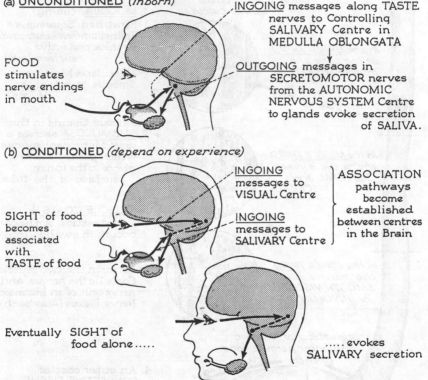

FOOD stimulates nerve endings in mouth

<u>INGOING</u> messages along TASTE nerves to Controlling SALIVARY Centre in MEDULLA OBLONGATA

<u>OUTGOING</u> messages in SECRETOMOTOR nerves from the AUTONOMIC NERVOUS SYSTEM Centre to glands evoke secretion of SALIVA.

(b) <u>CONDITIONED</u> *(depend on experience)*

SIGHT of food becomes associated with TASTE of food

<u>INGOING</u> messages to VISUAL Centre

<u>INGOING</u> messages to SALIVARY Centre

ASSOCIATION pathways become established between centres in the Brain

Eventually SIGHT of food alone.....

..... evokes SALIVARY secretion

Similar conditioned reflexes are established by smell, by thought of food, and even by the sounds of its preparation.

3

OESOPHAGUS

The Oesophagus is a muscular tube about 25 cm. long which conveys ingested food and fluid from the Mouth to the Stomach.

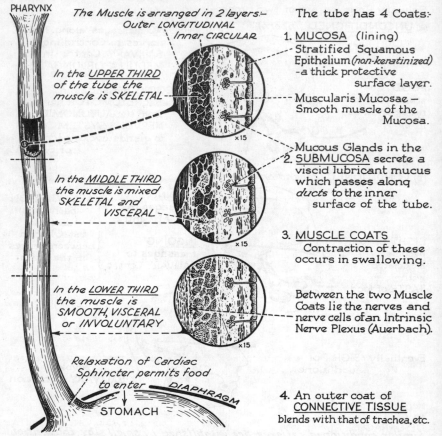

PHARYNX

The Muscle is arranged in 2 layers:-
Outer LONGITUDINAL
Inner CIRCULAR

In the _UPPER THIRD_ of the tube the muscle is SKELETAL

In the _MIDDLE THIRD_ the muscle is mixed SKELETAL and VISCERAL

In the _LOWER THIRD_ the muscle is SMOOTH, VISCERAL or INVOLUNTARY

Relaxation of Cardiac Sphincter permits food to enter
DIAPHRAGM
STOMACH

×15

×15

×15

The tube has 4 Coats:-

1. <u>MUCOSA</u> (lining)
- Stratified Squamous Epithelium _(non-keratinized)_ - a thick protective surface layer.
- Muscularis Mucosae — Smooth muscle of the Mucosa.

- Mucous Glands in the
2. <u>SUBMUCOSA</u> secrete a viscid lubricant mucus which passes along _ducts_ to the inner surface of the tube.

3. <u>MUSCLE COATS</u>
Contraction of these occurs in swallowing.

Between the two Muscle Coats lie the nerves and nerve cells of an Intrinsic Nerve Plexus (Auerbach).

4. An outer coat of <u>CONNECTIVE TISSUE</u> blends with that of trachea, etc.

Except during passage of food the oesophagus is flattened and closed; its mucosa thrown into several longitudinal folds.

SWALLOWING

Swallowing is a complex act initiated *voluntarily* and completed *involuntarily* (or *reflexly*)

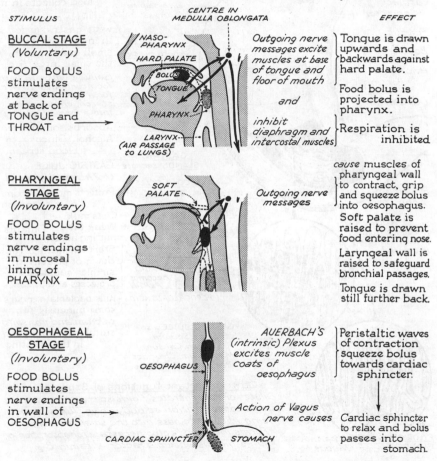

STIMULUS   CENTRE IN MEDULLA OBLONGATA   EFFECT

BUCCAL STAGE
(Voluntary)

FOOD BOLUS stimulates nerve endings at back of TONGUE and THROAT

NASO-PHARYNX
HARD, PALATE
BOLUS
TONGUE
PHARYNX
LARYNX
(AIR PASSAGE to LUNGS)

Outgoing nerve messages excite muscles at base of tongue and floor of mouth

and

inhibit diaphragm and intercostal muscles

Tongue is drawn upwards and backwards against hard palate.

Food bolus is projected into pharynx.

Respiration is inhibited.

PHARYNGEAL STAGE
(Involuntary)

FOOD BOLUS stimulates nerve endings in mucosal lining of PHARYNX

SOFT PALATE

Outgoing nerve messages

cause muscles of pharyngeal wall to contract, grip and squeeze bolus into oesophagus.
Soft palate is raised to prevent food entering nose.
Laryngeal wall is raised to safeguard bronchial passages.
Tongue is drawn still further back.

OESOPHAGEAL STAGE
(Involuntary)

FOOD BOLUS stimulates nerve endings in wall of OESOPHAGUS

OESOPHAGUS

CARDIAC SPHINCTER   STOMACH

AUERBACH'S (intrinsic) Plexus excites muscle coats of oesophagus

Action of Vagus nerve causes

Peristaltic waves of contraction squeeze bolus towards cardiac sphincter.

Cardiac sphincter to relax and bolus passes into stomach.

STOMACH

Oesophagus ----

Cardiac
Sphincter ----

Fundus.........

Body -

Pyloric
Portion

Duodenum

LESSER CURVATURE

GREATER CURVATURE

Pyloric
Sphincter

Rugae
[Irregular folds
when stomach
is empty]

Bulge in alimentary — RESERVOIR for food;
tract - shape varies   distends as swallowed
with individual   food collects in it.
and with degree
of fullness.   CHURN – mixes food

Its outer wall   with gastric juice,
consists of 3   then delivers it in
Smooth muscle   small quantities to
coats —   the small intestine.
 — Longitudinal
 — Circular   [Saliva swallowed with
 — Oblique   food continues to act
                                    for a time on cooked
                                    starch, changing it
MUCOSAL   into a form of sugar.]
LINING ——

                                  — Absorbs some Water,
                                       Alcohol, Glucose to
GASTRIC   blood stream.
GLANDS secrete - GASTRIC JUICE
                                      (1-2 litres/day) Acid.
----*MUCOUS CELLS*
secrete MUCIN — Protects mucosa from
                                     its own secretion.
PEPTIC CELLS *RENNIN* — Clots milk.   } in
secrete   *LIPASE* —— Weak fat-   } infants
ENZYMES   splitting action

PEPSINOGEN   } Start chemical break-
in presence of   } down of protein to
HYDROCHLORIC   } simpler substances –
ACID   } proteoses and peptones.
becomes
OXYNTIC *PEPSIN*
CELLS
secrete ½%
HYDROCHLORIC ACID — Kills bacteria; renders
                                              some minerals (e.g.
× 40   calcium and iron salts)
                                              suitable for absorption
Body   GASTRIC JUICE also   in the intestine.
Region   contains Sodium,
               Potassium, Calcium,
               Magnesium, Chlorides
               Phosphates, Sulphates.

× 40

Pyloric
Region

PYLORIC
GLANDS ——→ *Alkaline*
                     Mucus

Other important functions of Stomach
Dilutes or concentrates fluids so that they are the
same concentration as body's own fluids before
being allowed to pass into the small intestine.
  Forms Intrinsic Factor necessary for absorption of
Anti-Pernicious Anaemia Factor, Vitamin B$_{12}$.

GASTRIC JUICE

SECRETION of Gastric Juice is under 2 types of CONTROL:-

(a) NERVOUS - Messages are conveyed rapidly from Brain Centre by Nerve Fibres of the AUTONOMIC NERVOUS SYSTEM for *immediate* effect. e.g. stimulation of Parasympathetic (Vagus) nerves to GASTRIC GLANDS → secretion of HIGHLY ACID JUICE containing ENZYMES.

(b) HUMORAL - Message is CHEMICAL and carried in BLOOD STREAM for *slower* and *longer-lasting* control. [Note:- *Chemical messengers (Hormones) travel all over the body even though the message stimulates only one part — e.g. in this case the Gastric Glands.*]

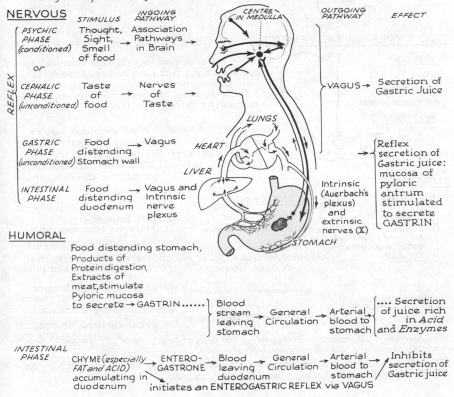

NERVOUS

	STIMULUS	INGOING PATHWAY		OUTGOING PATHWAY	EFFECT

REFLEX

PSYCHIC PHASE (conditioned) — Thought, Sight, Smell of food → Association Pathways in Brain

or

CEPHALIC PHASE (unconditioned) — Taste of food → Nerves of Taste

CENTRE IN MEDULLA

VAGUS → Secretion of Gastric Juice

GASTRIC PHASE (unconditioned) — Food distending Stomach wall → Vagus

INTESTINAL PHASE — Food distending duodenum → Vagus and Intrinsic nerve plexus

LUNGS
HEART
LIVER
STOMACH

Intrinsic (Auerbach's plexus) and extrinsic nerves (X)

→ Reflex secretion of Gastric juice: mucosa of pyloric antrum stimulated to secrete GASTRIN

HUMORAL

Food distending stomach, Products of Protein digestion, Extracts of meat, stimulate Pyloric mucosa to secrete → GASTRIN ...... Blood stream leaving stomach → General Circulation → Arterial blood to stomach → .... Secretion of juice rich in *Acid* and *Enzymes*

INTESTINAL PHASE

CHYME (especially FAT and ACID) accumulating in duodenum → ENTERO-GASTRONE → Blood leaving duodenum → General Circulation → Arterial blood to stomach → Inhibits secretion of Gastric juice

initiates an ENTEROGASTRIC REFLEX via VAGUS

MOVEMENTS of the STOMACH

Very little movement is seen in empty stomach until onset of HUNGER.

FILLING

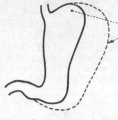

FUNDUS and
GREATER CURVATURE bulge and lengthen
as stomach fills with food. This is
"Receptive Relaxation" – the smooth muscle
cells relax so that the pressure in the
stomach does not rise until the organ is
fairly large.

TWO TYPES of MOVEMENT   occur while food is in stomach.

TONUS

'TONE RING'
with tone waves

PERISTALSIS

PYLORIC
SPHINCTER

HOPPER

MILL

BODY – Food lies in layers and the walls
exert a slight but steady pressure on it.
This squeezes food steadily towards the
PYLORUS even when stomach is becoming
relatively empty near the end of a meal.

From INCISURA ANGULARIS vigorous waves
of contraction mix food with digestive
juices and carry chyme through the
normally relaxed PYLORUS into the first
part of the duodenum.

EMPTYING

CONTROL of MOTILITY and EMPTYING

1. Enterogastric
   <u>NERVOUS</u> Reflex   *inhibition of VAGUS* →
       As Chyme enters and
       distends duodenum
   <u>HUMORAL</u>   *release of ENTEROGASTRONE* →
                                      to Blood Stream

> Inhibits tone and
>   peristalsis temporarily.
> Gastric motility depressed.
> Temporarily slows emptying
>   of stomach.

2.
       As duodenum   *stimulation of VAGUS* →
            empties
                    withdrawal of ENTEROGASTRONE →

> Waves of peristalsis become
>   stronger.
> Gastric motility increased.
> Emptying speeded up again.

*This mechanism is important in regulating the rate of emptying of the stomach from
moment to moment during the digestion of a meal.*

*[STARCHY FOODS leave the stomach quickly: MEAT leaves relatively slowly:
FATTY FOODS pass through most slowly of all.]*

34

PANCREAS

The Pancreas is a large gland lying across the Posterior Abdominal Wall. It has 2 secretions – a *digestive* secretion poured into the duodenum, and a *hormonal* secretion passed into the blood stream.

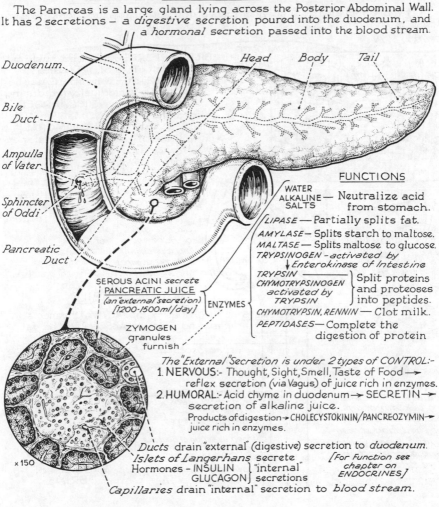

Duodenum

Bile Duct

Ampulla of Vater

Sphincter of Oddi

Pancreatic Duct

Head   Body   Tail

SEROUS ACINI *secrete*
PANCREATIC JUICE
(an "external" secretion)
[1200-1500ml/day]

ZYMOGEN granules furnish

ENZYMES

x 150

FUNCTIONS

WATER
ALKALINE — Neutralize acid
SALTS   from stomach.

LIPASE — Partially splits fat.

AMYLASE — Splits starch to maltose.

MALTASE — Splits maltose to glucose.

TRYPSINOGEN - activated by
     ↓ *Enterokinase of Intestine*

TRYPSIN ——————
CHYMOTRYPSINOGEN   } Split proteins
activated by   and proteoses
TRYPSIN   into peptides.

CHYMOTRYPSIN, RENNIN — Clot milk.

PEPTIDASES — Complete the
     digestion of protein

The "External" Secretion is under 2 types of CONTROL:-
1. NERVOUS:- Thought, Sight, Smell, Taste of Food →
     reflex secretion (via Vagus) of juice rich in enzymes.
2. HUMORAL:- Acid chyme in duodenum → SECRETIN →
     secretion of alkaline juice.
     Products of digestion → CHOLECYSTOKININ/PANCREOZYMIN →
     juice rich in enzymes.

Ducts drain "external" (digestive) secretion to *duodenum*.
Islets of Langerhans secrete
Hormones - INSULIN   } "internal"   [For Function see
         GLUCAGON } secretions   chapter on
                           ENDOCRINES]
Capillaries drain "internal" secretion to *blood stream*.

LIVER (and GALL BLADDER)

The Liver is a large, highly complex organ with many functions. One of these is to secrete 500-1000 ml. of BILE per day.

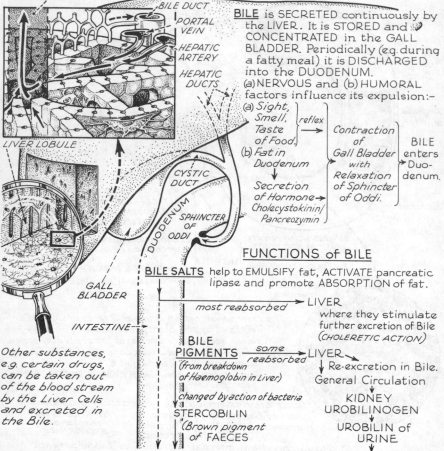

CENTRAL VEIN

BILE DUCT
PORTAL VEIN
HEPATIC ARTERY
HEPATIC DUCTS

LIVER LOBULE

CYSTIC DUCT
DUODENUM
SPHINCTER OF ODDI

GALL BLADDER

INTESTINE

BILE is SECRETED continuously by the LIVER. It is STORED and CONCENTRATED in the GALL BLADDER. Periodically (e.g. during a fatty meal) it is DISCHARGED into the DUODENUM.
(a) NERVOUS and (b) HUMORAL factors influence its expulsion:-

(a) Sight, Smell, Taste of Food. → reflex → Contraction of Gall Bladder with Relaxation of Sphincter of Oddi. → BILE enters Duodenum.

(b) Fat in Duodenum → Secretion of Hormone → Cholecystokinin / Pancreozymin →

FUNCTIONS of BILE

BILE SALTS help to EMULSIFY fat, ACTIVATE pancreatic lipase and promote ABSORPTION of fat.

most reabsorbed → LIVER where they stimulate further excretion of Bile (CHOLERETIC ACTION)

BILE PIGMENTS — some reabsorbed → LIVER
(from breakdown of Haemoglobin in Liver)
→ Re-excretion in Bile.
General Circulation
↓
KIDNEY
UROBILINOGEN
↓
UROBILIN of URINE

changed by action of bacteria
STERCOBILIN
(Brown pigment of FAECES

Other substances, e.g. certain drugs, can be taken out of the blood stream by the Liver Cells and excreted in the Bile.

36

SMALL INTESTINE

The Small Intestine is a long muscular tube – over 6 metres in length.
It receives CHYME in small quantities from the STOMACH; PANCREATIC
JUICE from the PANCREAS; BILE from the GALL BLADDER.

DUODENUM (10"-12")

It has 4 Coats:-
1. Outer <u>SEROUS COAT</u> of Peritoneum _(with vessels & nerves)._
2. <u>MUSCULAR COAT</u>
   Smooth muscle – 2 layers:-
   Outer- LONGITUDINAL
   Inner- CIRCULAR

STOMACH

3. <u>SUBMUCOUS</u>
   <u>COAT</u> with b.v's,
   fibrous tissue
   and (in duo-
   denum only)
   Brunner's glands-
   →alkaline juice and mucus
   protect mucosa from
   gastric juice

JEJUNUM (8')

4. <u>MUCOUS COAT</u>
   (or lining)

× 10

Gradual
transition
to

ILEUM (12')

× 10

VILLI-
finger-like projections (microvilli)
– increase surface area for
ABSORPTION

ILEO-
CAECAL
VALVE

20 to 30
Aggregations of
Lymph follicles in
PEYER'S PATCH _form part of_
the ileum's defence mechanism
against bacteria.

× 10

<u>MOVEMENTS</u> of the wall mix
food with digestive juices,
promote absorption and move
the residue along the tube.

Between the muscular layers
lies AUERBACH'S Nerve Plexus
through which peristaltic
movements are controlled.

<u>CRYPTS of LIEBERKÜHN</u>
secrete alkaline Intestinal
Juice (Succus Entericus).
PEPTIDASES – Split peptides to
                amino acids.
ENTEROKINASE- Activates
        Pancreatic Trypsinogen.
AMYLASE- Splits Starch to
                Maltose.
MALTASE ⎞ Complete digestion
LACTASE ⎬ of carbohydrates
SUCRASE ⎠ to simple sugars.
LIPASE- Splits fats to lower
        glycerides, fatty acids
        and glycerol.
   Recent work suggests that
enzymes in the Succus
Entericus are probably con-
tained in shed epithelial cells.

   The MUCOSA also secretes
HORMONES –
e.g. ENTEROGASTRONE,
     SECRETIN,
     CHOLECYSTOKININ/
          PANCREOZYMIN —
to help regulate flow of
Gastro-intestinal secretions.

MOVEMENTS of the SMALL INTESTINE

The Duodenum receives food in small quantities from the Stomach. The mixture of food and digestive juices — Chyme — is passed along the length of the Small Intestine.

TWO TYPES of MOVEMENT

SEGMENTATION —
Rhythmical alternating contractions and relaxations

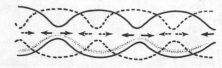

— These "shuttling" movements serve to mix CHYME and to bring it into contact with the absorptive mucosa, i.e. DIGESTION and ABSORPTION are promoted.

This type of movement is MYOGENIC, i.e. it is the property of the smooth muscle cells. It does not depend on a nervous mechanism.

PERISTALSIS —
Food probably acts as stimulus to stretch receptors in muscle and perhaps in mucous membrane

Circular muscle behind bolus *CONTRACTS*   Muscle in front of bolus *RELAXES*

— Waves of this contraction move the food along the canal.

The contraction behind the bolus sweeps it into the relaxed portion of the tube ahead.

This type of movement is NEUROGENIC, i.e. it is carried out reflexly through the nerve plexuses in the wall of the tube. The extrinsic nerves influence it. Parasymp. stimulation increases motility: Sympathetic decreases it.

EMPTYING

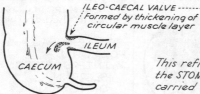

ILEO-CAECAL VALVE ------
Formed by thickening of circular muscle layer

ILEUM

CAECUM

opens and closes during digestion to allow spurts of fluid material from the Ileum to enter the Large Intestine.

This reflex is initiated when food enters the STOMACH. Motor nerve messages are carried to the wall of the tube by the autonomic nerves. (see page 149)

Meals of different composition travel along the intestine at different rates. DIGESTION and ABSORPTION of food are usually complete by the time the residue reaches the Ileo-caecal valve.

ABSORPTION in SMALL INTESTINE

Absorption of most digested foodstuffs occurs in the Small Intestine through the Striated Border Epithelium covering the Villi.

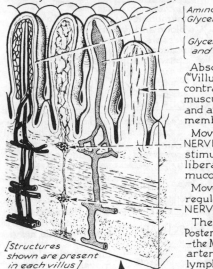

Amino Acids; Sugars; Minerals; Glycerol; some Fatty Acids and Vitamins	into CAPILLARIES
Glycerides and some Fatty Acids and Fat-soluble Vitamins	into LACTEALS

Absorption is aided by movements of Villi ("Villus Pump" mechanism) brought about by contraction of smooth muscle–extensions of muscularis mucosae– present in core of villus and attached to lacteal and to basement membrane of epithelium.

Movements of the villi are regulated through NERVE PLEXUS of MEISSNER and perhaps stimulated by intestinal hormones liberated to Blood Stream by Duodenal mucosa.

Movements of the muscular coats are regulated through the NERVE PLEXUS of AUERBACH.

The Intestines are suspended from the Posterior Abdominal wall by a delicate membrane –the MESENTERY–which carries the mesenteric arteries and veins to and from the gut; and lymphatic vessels on their way via the mesenteric lymph nodes to Thoracic Duct.

[Structures shown are present in each villus]

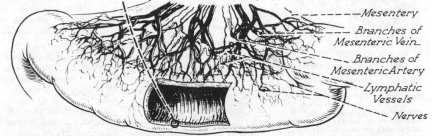

— — — — — Mesentery

— — — Branches of Mesenteric Vein

— — — Branches of Mesenteric Artery

— Lymphatic Vessels

— — Nerves

Absorption is not just a process of simple diffusion of substances from areas of high to areas of low concentration. Movement of ions can take place against a concentration gradient. In other words— Absorption is often an active process involving the use of energy by the epithelial cells.

39

LARGE INTESTINE

(Total length over 6')

Mesentery

SEROUS COAT
Peritoneum - moist
membrane - carries
blood and lymph
vessels and nerves;
reduces friction
between surfaces

Splenic
Flexure

Hepatic
Flexure

Transverse Colon

Taenia

Ascending
Colon

MUCOUS COAT

*Striated border
epithelium*

*Capillary
bloodvessels*

Goblet cells

Lymph nodule

**SUBMUCOUS
COAT**

**MUSCULAR
COAT**

*CIRCULAR
COAT*

*INCOMPLETE
LONGITUDINAL COAT* -
3 tape-like
bands

Smooth
muscle

×20

*Auerbach's
Nerve Plexus*

Caecum

--Vermiform
Appendix

Ileo-caecal valve

Sigmoid
Flexure

Pelvic Colon

Rectum

Anal Canal

Internal
(SMOOTH CIRC. M.) &
External
(VOLUNTARY MUSC.)
Anal Sphincters

SKIN

FUNCTIONS

Absorption of Water
and Salts to blood-
vessels in colon wall
to conserve body's
fluid and to dry faeces.

Mucus to lubricate
and neutralize faeces.
Defence against
bacteria.

Loose fibrous tissue
with blood vessels,
lymph vessels, nerves.

Mass peristalsis
moves the contents
towards the rectum.

Storage of faeces
until defaecation.

*[In large intestine
Bacteria synthesize
some vitamins
e.g. B and K]*

Contractions of
muscle coats expel
faeces from body.

Relaxation
permits
defaecation.

FAECES: About 100 grams/day
(amount varies with diet)
Consist mainly of:- Residues of indigestible
material in Food (e.g. skins of Fruits),
Bile (Pigments and Salts),
Intestinal secretions (including mucus),
Leucocytes (migrate from Blood Stream),
Shed epithelial cells,
Large numbers of Bacteria (make
up to ⅓ of total solids),
Inorganic material (10-20%)
chiefly calcium and phosphates,
Very little digestible food is present.

MOVEMENTS of the LARGE INTESTINE

The Ileo-caecal Valve opens and closes during digestion. Peristaltic waves sweep semi-fluid contents of Ileum through the relaxed valve. During its stay in the Large Intestine faecal matter is subjected to –

TWO TYPES OF MOVEMENT

"TONE WAVES"
run forwards –
and backwards from
Cannon's Point to the
Ileum – (sometimes
mistakenly called
"Anti-peristalsis")
These waves "die out"

CANNON'S POINT

Ensure prolonged contact
of contents with mucosa
— and promote absorption
of Water and Salt from
faeces –
This type of movement is
MYOGENIC

MASS PERISTALSIS
Strong waves
at infrequent
intervals start
at upper end of
Ascending Colon

Empty Transverse Colon
and sweep faeces into the
Descending and Pelvic
Colons and into Rectum –
— **NEUROGENIC**

[Reflex often initiated by
passage of food into Stomach
(Gastro-colic Reflex)]

EMPTYING
(DEFAECATION)

Complex
REFLEX act

Stimulus
Passage of
faeces into
the Rectum
distends
wall

[
+
passage
through
anal
canal
]

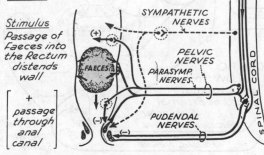

SYMPATHETIC
NERVES

FAECES

PELVIC
NERVES

PARASYMP.
NERVES

PUDENDAL
NERVES

SPINAL CORD

Sensations rise to level of consciousness →
Voluntary decision → Impulses to
inhibit or permit reflex evacuation

Outgoing nerve messages →
Powerful peristaltic
contractions of Descending
Colon, Pelvic Colon and Rectum

[Preceded by inspiratory
descent of diaphragm and
voluntary contraction of
abdominal muscles to raise
intra-abdominal pressure]

Contraction of Pelvic Floor
muscles with Relaxation of
Anal Sphincter
↓
Evacuation of Faeces

NERVOUS CONTROL of GUT MOVEMENTS

The movements of the Alimentary Canal are carried out automatically and, for the most part, beneath the level of consciousness.

Movements in the wall of the Alimentary Canal are
either ⟨ *MYOGENIC* — *a property of the smooth muscle.*
NEUROGENIC — *dependent on the Intrinsic Nerve Plexuses.*

They can occur even after *Extrinsic* nerves to the tract have been cut. Normally, however, impulses travelling in these nerves of the SYMPATHETIC and PARASYMPATHETIC Systems, from the CONTROLLING CENTRES of the AUTONOMIC NERVOUS SYSTEM in the BRAIN and SPINAL CORD, *influence* and *co-ordinate* events in the whole tract.

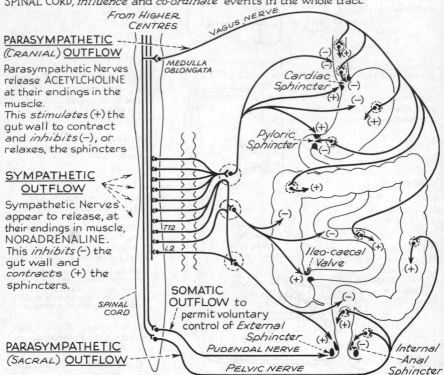

From Higher Centres

VAGUS NERVE

PARASYMPATHETIC (*CRANIAL*) OUTFLOW

MEDULLA OBLONGATA

Parasympathetic Nerves release ACETYLCHOLINE at their endings in the muscle.
This *stimulates* (+) the gut wall to contract and *inhibits* (−), or relaxes, the sphincters

Cardiac Sphincter

Pyloric Sphincter

SYMPATHETIC OUTFLOW

Sympathetic Nerves appear to release, at their endings in muscle, NORADRENALINE.
This *inhibits* (−) the gut wall and *contracts* (+) the sphincters.

T12
L2

Ileo-caecal Valve

SPINAL CORD

SOMATIC OUTFLOW to permit voluntary control of *External Sphincter*
PUDENDAL NERVE

PARASYMPATHETIC (*SACRAL*) OUTFLOW

PELVIC NERVE

Internal Anal Sphincter

TRANSPORT and UTILIZATION of FOODSTUFFS

After absorption the NUTRIENTS are transported in:-

(a) BLOOD
in Mesenteric Veins
to PORTAL VEIN
to Liver

(b) LYMPH
in Lymphatic Vessels to
THORACIC DUCT and via
a large vein in the neck to
Systemic Circulation

↓

which then distributes FOOD
to ALL TISSUES of the BODY
for UTILIZATION :-
PROTEIN → To build or repair
protoplasm.
CARBOHYDRATES → For oxidation
with release of Energy
for work or as heat to
maintain body temperature.
FAT → For storage in fat depots.

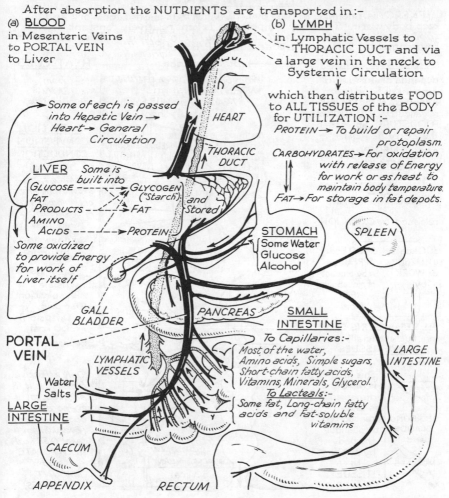

Some of each is passed
into Hepatic Vein →
Heart → General
Circulation

HEART

THORACIC
DUCT

LIVER   Some is
built into
GLUCOSE - - - ➤ GLYCOGEN
FAT   ("Starch")
PRODUCTS - - - ➤ FAT   and
AMINO   Stored
ACIDS - - - ➤ PROTEIN

Some oxidized
to provide Energy
for work of
Liver itself

STOMACH
Some Water
Glucose
Alcohol

SPLEEN

GALL
BLADDER

PANCREAS

SMALL
INTESTINE
To Capillaries:-
Most of the water,
Amino acids, Simple sugars,
Short-chain fatty acids,
Vitamins, Minerals, Glycerol.
To Lacteals:-
Some fat, Long-chain fatty
acids and fat-soluble
vitamins

PORTAL
VEIN

LYMPHATIC
VESSELS

LARGE
INTESTINE

Water
Salts

LARGE
INTESTINE

CAECUM

APPENDIX   RECTUM

43

HEAT BALANCE

ENERGY is released in cells by OXIDATION. It is used for work and to keep body warm. The Body Temperature is kept relatively constant *(with a slight fluctuation throughout the 24 hours)* in spite of wide variations in Environmental Temperature and Heat Production.

HEAT PRODUCTION ———— *must balance* ———— HEAT LOSS

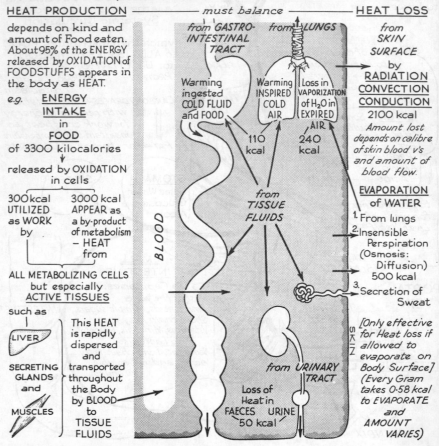

depends on kind and amount of Food eaten. About 95% of the ENERGY released by OXIDATION of FOODSTUFFS appears in the body as HEAT.

e.g. **ENERGY INTAKE** in **FOOD**

of 3300 kilocalories

↓

released by OXIDATION in cells

300 kcal UTILIZED as WORK by

3000 kcal APPEAR as a by-product of metabolism – HEAT from

ALL METABOLIZING CELLS but especially **ACTIVE TISSUES**

such as

LIVER

SECRETING GLANDS and

MUSCLES

This HEAT is rapidly dispersed and transported throughout the Body by BLOOD to TISSUE FLUIDS

from GASTRO-INTESTINAL TRACT

Warming ingested COLD FLUID and FOOD

from LUNGS

Warming INSPIRED COLD AIR

Loss in VAPORIZATION of H_2O in EXPIRED AIR

110 kcal

240 kcal

from TISSUE FLUIDS

BLOOD

from URINARY TRACT

Loss of Heat in FAECES URINE 50 kcal

from **SKIN SURFACE** by **RADIATION CONVECTION CONDUCTION** 2100 kcal *Amount lost depends on calibre of skin blood v's and amount of blood flow.*

EVAPORATION of WATER

1. From lungs

2. Insensible Perspiration (Osmosis: Diffusion) 500 kcal

3. Secretion of Sweat

[Only effective for Heat loss if allowed to evaporate on Body Surface] *(Every Gram takes 0·58 kcal to EVAPORATE and AMOUNT VARIES)*

SKIN

MAINTENANCE of BODY TEMPERATURE

Any tendency for the BODY TEMPERATURE to *RISE*
as by
1. INCREASED HEAT PRODUCTION — is balanced by INCREASED HEAT LOSS

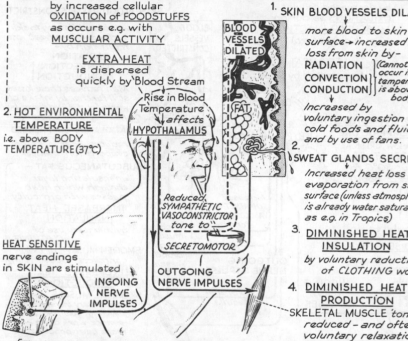

by increased cellular
OXIDATION of FOODSTUFFS
as occurs e.g. with
MUSCULAR ACTIVITY

EXTRA HEAT
is dispersed
quickly by Blood Stream

Rise in Blood
Temperature
affects
HYPOTHALAMUS

2. HOT ENVIRONMENTAL
TEMPERATURE
i.e. above BODY
TEMPERATURE (37°C)

BLOOD
VESSELS
DILATED

FAT

Reduced
SYMPATHETIC
VASOCONSTRICTOR
tone to
SECRETOMOTOR

HEAT SENSITIVE
nerve endings
in SKIN are stimulated

INGOING
NERVE
IMPULSES

OUTGOING
NERVE IMPULSES

1. SKIN BLOOD VESSELS DILATE

more blood to skin
surface → increased heat
loss from skin by -
RADIATION   (Cannot
CONVECTION   occur if air
CONDUCTION   temperature
              is above
              body's)
Increased by
voluntary ingestion of
cold foods and fluids
and by use of fans.

2. SWEAT GLANDS SECRETE

Increased heat loss by
evaporation from skin
surface (unless atmosphere
is already water saturated
as e.g. in Tropics)

3. DIMINISHED HEAT
   INSULATION
   by voluntary reduction
   of CLOTHING worn

4. DIMINISHED HEAT
   PRODUCTION
   SKELETAL MUSCLE 'tone'
   reduced - and often
   voluntary relaxation
   → less work done →
   less heat produced

5. REDUCTION of 'ENERGY INTAKE'
   by voluntary restriction of
   PROTEIN in DIET

[The increase in activity during the day
probably accounts for the gradual Physiological
rise in body temperature from about 96.5°F
(35.8°C) in the early morning to about 99.2°F
(37.3°C) in the late afternoon.

Unless exercise is very strenuous or environ-
ment is very hot and humid these measures → RESTORE BODY TEMPERATURE
to NORMAL

4

MAINTENANCE of BODY TEMPERATURE

Any tendency for the BODY TEMPERATURE to *FALL*
as by

1. **DIMINISHED HEAT PRODUCTION** — is balanced by **DIMINISHED HEAT LOSS**

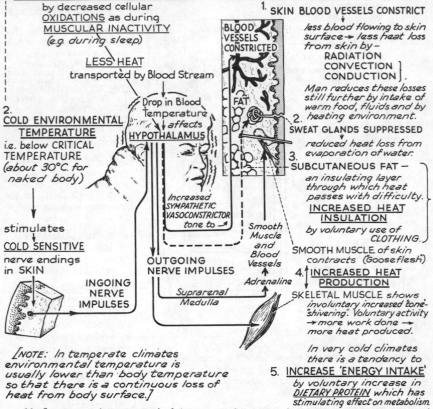

by decreased cellular
OXIDATIONS as during
MUSCULAR INACTIVITY
(e.g. during sleep)

LESS HEAT
transported by Blood Stream

2.
**COLD ENVIRONMENTAL
TEMPERATURE**
i.e. below CRITICAL
TEMPERATURE
*(about 30°C. for
naked body)*

stimulates

COLD SENSITIVE
nerve endings
in SKIN

Drop in Blood
Temperature
affects
HYPOTHALAMUS

*Increased
SYMPATHETIC
VASOCONSTRICTOR
tone to* ➝

**OUTGOING
NERVE IMPULSES**

**INGOING
NERVE
IMPULSES**

*Sup.rarenal
Medulla*

1. **SKIN BLOOD VESSELS CONSTRICT**

BLOOD
VESSELS
CONSTRICTED

FAT

*less blood flowing to skin
surface* ➝ *less heat loss
from skin by –*
RADIATION
CONVECTION }
CONDUCTION

*Man reduces these losses
still further by intake of
warm food, fluids and by
heating environment.*

2. **SWEAT GLANDS SUPPRESSED**

*reduced heat loss from
evaporation of water.*

3. **SUBCUTANEOUS FAT –**
*an insulating layer
through which heat
passes with difficulty.*
**INCREASED HEAT
INSULATION**
*by voluntary use of
CLOTHING.*

SMOOTH MUSCLE *of skin
contracts* (Gooseflesh)

*Smooth
Muscle
and
Blood
Vessels*

Adrenaline

4. **INCREASED HEAT
PRODUCTION**

SKELETAL MUSCLE *shows
involuntary increased tone-
"shivering". Voluntary activity
➝ more work done ➝
more heat produced.*

*In very cold climates
there is a tendency to*

5. **INCREASE 'ENERGY INTAKE'**
*by voluntary increase in
DIETARY PROTEIN which has
stimulating effect on metabolism.*

*[NOTE: In temperate climates
environmental temperature is
usually lower than body temperature
so that there is a continuous loss of
heat from body surface.]*

*Unless environmental temperature
is very low these measures tend to* — **RESTORE BODY TEMPERATURE**
to NORMAL

CHAPTER 3.
TRANSPORT SYSTEM

All cells are bathed by TISSUE FLUID. It is from solution in this fluid that O_2 and food materials diffuse into each cell; to it waste products including CO_2 diffuse out of the cells.

The CARDIOVASCULAR SYSTEM is the TRANSPORT SYSTEM which conveys these materials to and from the tissues.

This simplified diagram gives a concept of the general plan of the circulation.

CAPILLARY BED-

HEAD, NECK, UPPER LIMBS

*__PULMONARY__
(or lesser) __CIRC__ⁿ—
from Right Vent.
to Left Auricle*

O_2 is taken up by blood and CO_2 given off

PULMONARY CIRCULATION

R.A.   L.A.

R.V.   L.V.

*__SYSTEMIC__
(or greater) __CIRC__ⁿ—
from Left Ventricle
to Right Auricle*

HEPATIC or PORTAL CIRCⁿ

Food is absorbed

PORTAL VEIN

CAPILLARY BED- DIGESTIVE TRACT

Excess water and waste products are filtered off

RENAL CIRCULATION

CAPILLARY BED- TRUNK, LOWER LIMBS

HEART....*a "Pump"* which drives —

—BLOOD.... *a complex fluid* containing food materials, respiratory gases, waste products, protective and regulating chemical substances round —

—BLOOD VESSELS....*a closed system of tubes:*—

ARTERIES....from the *"pump"* to the tissues of the body.

branch into

CAPILLARIES....where gases, food and waste substances pass — from/to the Blood

from/to the Tissue Fluids.

reunite to form

VEINS....from the tissues of the body back to the *"pump".*

47

HEART

The Heart has 4 CHAMBERS - *Lined with Endothelium*
ENDOCARDIUM
Thick walls of Cardiac Muscle —
MYOCARDIUM
Enclosed in a 2-layered serous membrane —
PERICARDIUM
A thin film of fluid separates the 2 layers of Pericardial Sac

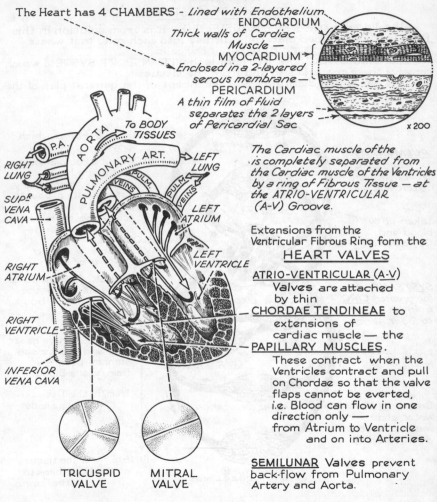

x 200

The Cardiac muscle of the is completely separated from the Cardiac muscle of the Ventricles by a ring of Fibrous Tissue — at the ATRIO-VENTRICULAR (A-V) Groove.

Extensions from the Ventricular Fibrous Ring form the

HEART VALVES

ATRIO-VENTRICULAR (A-V)
Valves are attached by thin
CHORDAE TENDINEAE to
extensions of cardiac muscle — the
PAPILLARY MUSCLES.
These contract when the Ventricles contract and pull on Chordae so that the valve flaps cannot be everted, i.e. Blood can flow in one direction only — from Atrium to Ventricle and on into Arteries.

SEMILUNAR Valves prevent back-flow from Pulmonary Artery and Aorta.

AORTA
To BODY TISSUES
P.A.
PULMONARY ART.
RIGHT LUNG
LEFT LUNG
VEINS PULM.
PULM. VEINS
SUP.ᴿ VENA CAVA
LEFT ATRIUM
RIGHT ATRIUM
LEFT VENTRICLE
RIGHT VENTRICLE
INFERIOR VENA CAVA

TRICUSPID VALVE

MITRAL VALVE

48

HEART

The human heart is really a *DOUBLE PUMP* — each quite separate from the other

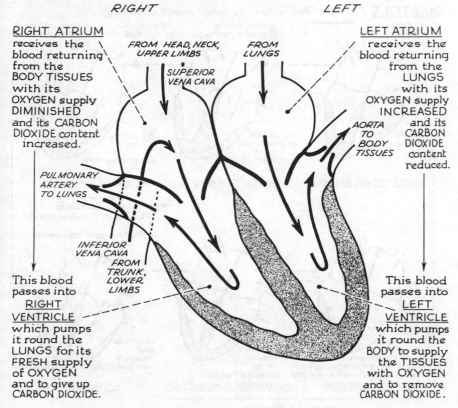

RIGHT   LEFT

<u>RIGHT ATRIUM</u> receives the blood returning from the BODY TISSUES with its OXYGEN supply DIMINISHED and its CARBON DIOXIDE content increased.

FROM HEAD, NECK, UPPER LIMBS

SUPERIOR VENA CAVA

FROM LUNGS

<u>LEFT ATRIUM</u> receives the blood returning from the LUNGS with its OXYGEN supply INCREASED and its CARBON DIOXIDE content reduced.

AORTA TO BODY TISSUES

PULMONARY ARTERY TO LUNGS

INFERIOR VENA CAVA

FROM TRUNK, LOWER LIMBS

This blood passes into <u>RIGHT VENTRICLE</u> which pumps it round the LUNGS for its FRESH supply of OXYGEN and to give up CARBON DIOXIDE.

This blood passes into <u>LEFT VENTRICLE</u> which pumps it round the BODY to supply the TISSUES with OXYGEN and to remove CARBON DIOXIDE.

This diagram simplifies the structure of the heart to make it easier to understand the function of its various parts.

CARDIAC CYCLE

Diagrammatic representation of the sequence of events in the heart during ONE heart beat.

DIASTOLE [Period of Relaxation – i.e. when heart is resting]

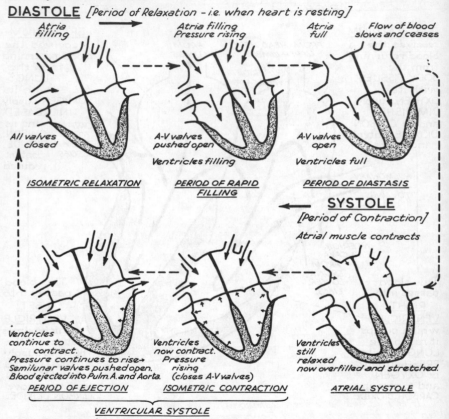

Atria filling

Atria filling
Pressure rising

Atria full

Flow of blood slows and ceases

All valves closed

A-V valves pushed open

Ventricles filling

A-V valves open

Ventricles full

ISOMETRIC RELAXATION

PERIOD OF RAPID FILLING

PERIOD OF DIASTASIS

SYSTOLE
[Period of Contraction]

Atrial muscle contracts

Ventricles continue to contract.
Pressure continues to rise→
Semilunar valves pushed open.
Blood ejected into Pulm. A. and Aorta.

Ventricles now contract.
Pressure rising
(closes A-V valves)

Ventricles still relaxed now overfilled and stretched.

PERIOD OF EJECTION

ISOMETRIC CONTRACTION

ATRIAL SYSTOLE

VENTRICULAR SYSTOLE

The total cycle of events takes about 0·8 second when heart is beating 75 times per minute.

HEART SOUNDS

During each CARDIAC CYCLE 2 HEART SOUNDS can be heard through a STETHOSCOPE applied to the CHEST WALL.

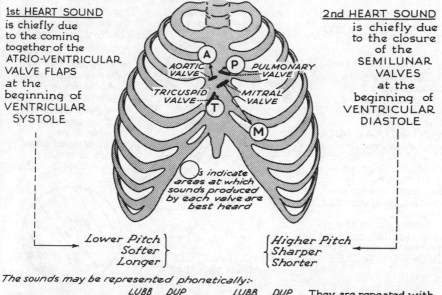

1st HEART SOUND
is chiefly due
to the coming
together of the
ATRIO-VENTRICULAR
VALVE FLAPS
at the
beginning of
VENTRICULAR
SYSTOLE

AORTIC VALVE
TRICUSPID VALVE
PULMONARY VALVE
MITRAL VALVE

is indicate areas at which sounds produced by each valve are best heard

2nd HEART SOUND
is chiefly due
to the closure
of the
SEMILUNAR
VALVES
at the
beginning of
VENTRICULAR
DIASTOLE

Lower Pitch
Softer
Longer

Higher Pitch
Sharper
Shorter

The sounds may be represented phonetically:-

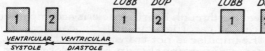

LUBB   DUP   LUBB   DUP
1   2   1   2   1   2

VENTRICULAR   VENTRICULAR
SYSTOLE   DIASTOLE

They are repeated with every CARDIAC CYCLE i.e. about 70 times per minute in the average healthy adult.

If the valves have been damaged by disease additional murmurs can be heard as the blood flows forwards through narrowed valves or leaks backwards through incompetent valves.

Occasionally additional sounds are heard over normal healthy hearts.

ORIGIN and CONDUCTION of HEART BEAT.

The rhythmic contraction of the heart is called the HEART BEAT. The impulse to contract is generated in specialized NODAL TISSUE in the wall of the RIGHT ATRIUM.

Impulses are discharged rhythmically from this SINO-ATRIAL NODE *(The 'Pacemaker')*

The wave of excitation spreads throughout the muscle of both ATRIA ······

which are then excited to contract

The Right Atrium starts contracting just before Left Atrium.

The impulse is picked up by another mass of NODAL TISSUE — the ATRIO-VENTRICULAR NODE and relayed by PURKINJE TISSUE — *(in Bundle of His and its branches)* lying beneath endocardium on the interventricular septum.

This relays the impulse to contract to the muscle of both VENTRICLES ············ which then contract together

A ring of fibrous tissue separates Atria from Ventricles. The heart beat is not transmitted from Atria to Ventricles directly by ordinary CARDIAC muscle.

x 150

The wave of excitation spreading through heart muscle is accompanied by electrical changes. These can be recorded by an electrocardiograph.

In complete heart block the impulse is not transmitted to the Ventricles and they contract independently. In incomplete heart block every 2nd. or 3rd. impulse gets through.

Although the heart initiates its own impulse to contract its activity is finely adjusted, to meet the body's constantly changing needs, by nervous impulses discharged from Controlling Centres in the Brain and Spinal Cord. The Sympathetic nerves increase the rate and force of the heart beat: the Parasympathetic nerves slow heart and reduce force of contraction *(see pp. 149,150).*

BLOOD VESSELS

The system of tubes through which the heart pumps blood.

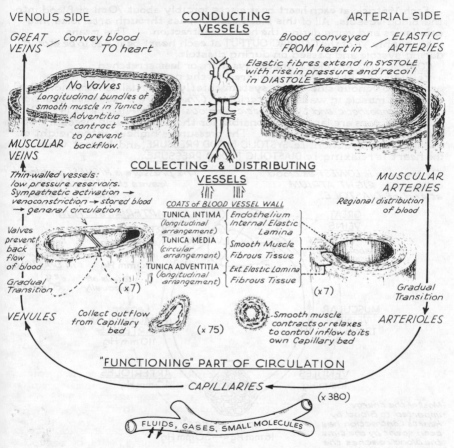

VENOUS SIDE

CONDUCTING VESSELS

ARTERIAL SIDE

GREAT VEINS — *Convey blood TO heart*

Blood conveyed FROM heart in — *ELASTIC ARTERIES*

Elastic fibres extend in SYSTOLE with rise in pressure and recoil in DIASTOLE

No Valves
Longitudinal bundles of smooth muscle in Tunica Adventitia contract to prevent backflow.

MUSCULAR VEINS

Thin-walled vessels: low pressure reservoirs. Sympathetic activation → venoconstriction → stored blood → general circulation.

COLLECTING & DISTRIBUTING VESSELS

MUSCULAR ARTERIES

Regional distribution of blood

COATS of BLOOD VESSEL WALL

TUNICA INTIMA (longitudinal arrangement)	Endothelium / Internal Elastic Lamina
TUNICA MEDIA (circular arrangement)	Smooth Muscle / Fibrous Tissue
TUNICA ADVENTITIA (longitudinal arrangement)	Ext. Elastic Lamina / Fibrous Tissue

Valves prevent back flow of blood

Gradual Transition

Gradual Transition

(x 7)

(x 7)

VENULES

Collect outflow from Capillary bed

(x 75)

Smooth muscle contracts or relaxes to control inflow to its own Capillary bed

ARTERIOLES

"FUNCTIONING" PART OF CIRCULATION

CAPILLARIES

(x 380)

FLUIDS, GASES, SMALL MOLECULES

Only from CAPILLARIES can Blood give up food and oxygen to tissues; and receive waste products and carbon dioxide from tissues.

BLOOD PRESSURE

Each Ventricle at each heart beat ejects forcibly about 70 ml. of Blood into the Blood Vessels. All of this blood cannot pass through arterioles into capillaries and veins during the heart's contraction. This means that roughly 5/8 of the <u>CARDIAC OUTPUT</u> at each heart beat has to be stored during systole and passed on during diastole.

<u>CONDUCTING ARTERIES</u> are always more or less stretched.

<u>PERIPHERAL RESISTANCE</u> is offered to the passage of blood from arterial to venous side of the system chiefly by partial constriction *("Tone")* of smooth muscle in walls of Arterioles. *(The calibre is regulated by action of Parasympathetic and Sympathetic Nervous System* — [see pages 149, 150]

These factors are largely responsible for the considerable pressure of the blood in the Arterial System. The pressure is highest at the height of the heart's contraction, i.e. <u>SYSTOLIC BLOOD PRESSURE,</u> and lowest when the heart is relaxing, i.e. <u>DIASTOLIC BLOOD PRESSURE</u>.

The pressure is LOWEST as blood drains into RIGHT ATRIUM at end of DIASTOLE.

The pressure is HIGHEST as blood leaves the LEFT VENTRICLE at end of SYSTOLE.

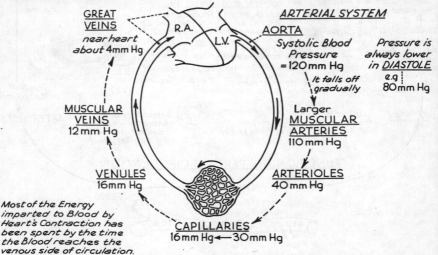

GREAT VEINS
near heart
about 4mm Hg

R.A.

L.V.

ARTERIAL SYSTEM

AORTA
Systolic Blood Pressure
= 120 mm Hg

Pressure is always lower in DIASTOLE
e.g
80 mm Hg

It falls off gradually

MUSCULAR VEINS
12 mm Hg

Larger MUSCULAR ARTERIES
110 mm Hg

VENULES
16mm Hg

ARTERIOLES
40 mm Hg

Most of the Energy imparted to Blood by Heart's Contraction has been spent by the time the Blood reaches the venous side of circulation.

CAPILLARIES
16mm Hg ← 30mm Hg

NOTE :- *Any alteration in the* <u>TOTAL AMOUNT</u> *or* <u>VISCOSITY</u> *of Blood will also affect BLOOD PRESSURE.*

MEASUREMENT of ARTERIAL BLOOD PRESSURE

The Arterial Blood Pressure is measured in man by means of a
SPHYGMOMANOMETER.

This consists of a
RUBBER BAG _(covered
with a cloth envelope)_
which is wrapped
round the UPPER ARM
over the BRACHIAL
ARTERY.

One tube connects the
inside of the bag with
a MANOMETER
containing MERCURY.

Another tube connects
the inside of the bag to
a hand operated PUMP
with a release VALVE.

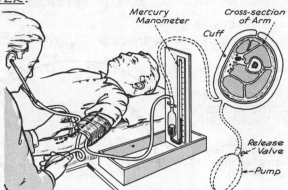

Mercury
Manometer

Cuff

Cross-section
of Arm

Release
Valve

Pump

METHOD

_Air is pumped into the rubber bag till
pressure in cuff is greater than pressure
in artery even during heart's systole —
           — Artery is then closed down during_ SYSTOLE and DIASTOLE
_[At same time air is pushing up mercury
           column in manometer.]_
_By releasing valve on pump the pressure
in cuff is gradually reduced till maximum
pressure in artery just overcomes pressure in
cuff — Some blood begins to spurt through during_ SYSTOLE — Artery still closed
                                                                during DIASTOLE

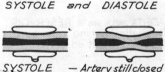

At this point FAINT rhythmical TAPPING SOUNDS
begin to be heard through STETHOSCOPE. The
height of mercury in millimetres is taken as the
SYSTOLIC Blood Pressure (e.g. 120 mm Hg)

_Pressure in cuff is reduced still further
till it is just less than the lowest pressure
in artery towards the end of diastole
(i.e. just before next heart beat) —
           — Blood flow is unimpeded during_ SYSTOLE and DIASTOLE.

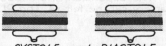

_The sounds stop. The height of mercury in the manometer at this point
is taken as the DIASTOLIC Blood Pressure (e.g. about 80 mm Hg)_

These values differ with SEX, AGE, EXERCISE, SLEEP, etc.

ELASTIC ARTERIES

The large <u>CONDUCTING ARTERIES</u> near the <u>HEART</u> are
<u>ELASTIC ARTERIES</u>

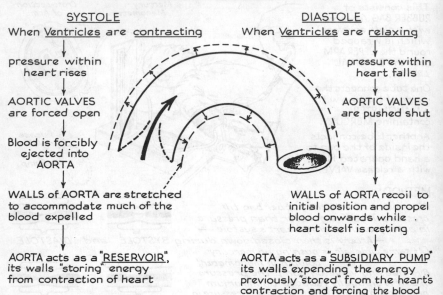

SYSTOLE
When <u>Ventricles</u> are <u>contracting</u>

↓

pressure within
heart rises

↓

AORTIC VALVES
are forced open

↓

Blood is forcibly
ejected into
AORTA

↓

WALLS of AORTA are stretched
to accommodate much of the
blood expelled

↓

AORTA acts as a <u>"RESERVOIR"</u>,
its walls "storing" energy
from contraction of heart

DIASTOLE
When <u>Ventricles</u> are <u>relaxing</u>

↓

pressure within
heart falls

↓

AORTIC VALVES
are pushed shut

↓

WALLS of AORTA recoil to
initial position and propel
blood onwards while
heart itself is resting

↓

AORTA acts as a <u>"SUBSIDIARY PUMP"</u>
its walls "expending" the energy
previously "stored" from the heart's
contraction and forcing the blood
on when the heart itself is resting.

As blood is pumped from the heart during systole, this distension
and increase in pressure which starts in the aorta passes along the
whole arterial system as a wave — the *pulse wave*.

The expansion and subsequent relaxation of the wall of the radial
artery can be felt as "The PULSE" at the wrist.

*A great increase in Blood Pressure can result if
these walls lose some of their elasticity with age or disease
and can no longer stretch readily to accommodate so much of
the heart's output during systole.*

CAPILLARIES : EXCHANGE of WATER & SOLUTES

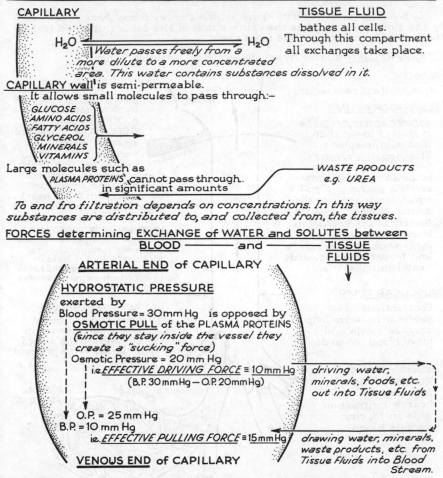

CAPILLARY

H_2O ⇌ H_2O

Water passes freely from a more dilute to a more concentrated area. This water contains substances dissolved in it.

CAPILLARY wall is semi-permeable.
It allows small molecules to pass through:-

GLUCOSE
AMINO ACIDS
FATTY ACIDS
GLYCEROL
MINERALS
VITAMINS

Large molecules such as
PLASMA PROTEINS cannot pass through.
in significant amounts

TISSUE FLUID

bathes all cells.
Through this compartment all exchanges take place.

WASTE PRODUCTS
e.g. UREA

To and fro filtration depends on concentrations. In this way substances are distributed to, and collected from, the tissues.

FORCES determining EXCHANGE of WATER and SOLUTES between
BLOOD ——— and ——— TISSUE
FLUIDS

ARTERIAL END of CAPILLARY

HYDROSTATIC PRESSURE
exerted by
Blood Pressure = 30mm Hg is opposed by
OSMOTIC PULL of the PLASMA PROTEINS
(since they stay inside the vessel they create a "sucking" force)
Osmotic Pressure = 20 mm Hg
i.e. *EFFECTIVE DRIVING FORCE* ≡ 10mm Hg
(B.P. 30 mm Hg — O.P. 20mm Hg)

O.P. = 25 mm Hg
B.P. = 10 mm Hg
i.e. *EFFECTIVE PULLING FORCE* ≡ 15mm Hg

VENOUS END of CAPILLARY

driving water, minerals, foods, etc. out into Tissue Fluids

drawing water, minerals, waste products, etc. from Tissue Fluids into Blood Stream.

These exchanges in Capillaries result in a continuous "turn-over" and renewal of Tissue Fluids.

VEINS: VENOUS RETURN

Capillaries unite to form *Veins* which *convey blood back to the heart.* By the time blood reaches the veins much of the force imparted to it by the heart's contraction has been spent.

<u>VENOUS RETURN</u> to the heart depends on various factors:-

<u>GRAVITY</u> -
   from head and neck region.

"RESPIRATORY PUMP"

(a) *Intrathoracic pressure* is always lower than that in atmosphere. This *"negative pressure"* (which fluctuates with Respiratory Movements) exerts a *"suctioning"* pull which tends to draw the column of blood upwards.

(b) Descent of *DIAPHRAGM* in Inspiration increases *Intra-abdominal pressure* and forces blood upwards in abdominal veins.

"MUSCULAR PUMP"

   Contractions of skeletal muscles help to squeeze veins and move blood upwards.

"CARDIAC PUMP"

   Residual force imparted by heart's contraction aids in forcing venous blood towards heart.

RIGHT ATRIUM

DIAPHRAGM

"<u>OVERFILLING</u>" of the system (i.e. adequate <u>BLOOD VOLUME</u>) and <u>ADEQUATE 'TONE'</u> of smooth muscle in venous walls are essential. (e.g. if too many vessels are dilated at any one time adequate venous return to the heart will not be maintained.)

   Sympathetic nerves keep veins partially constricted. Increased sympathetic 'tone' (e.g. after haemorrhage) mobilizes blood held in venous reservoirs and helps to maintain venous return to the heart.

<u>ACTION of VALVES</u>
Once the column of blood has moved up *UNIDIRECTIONAL VALVES* in larger veins prevent back flow.

CAPILLARIES

ARTERIAL PRESSURE

VENULES

WATER BALANCE

Water makes up about 70% of adult human body, i.e. about 46 litres in 70 kg. man.
Some is in Blood and Tissue Fluids; a great deal in cells themselves. In health the total
amount of body water (and salt) is kept reasonably constant in spite of wide
fluctuations in daily intake.

A BALANCE is struck between

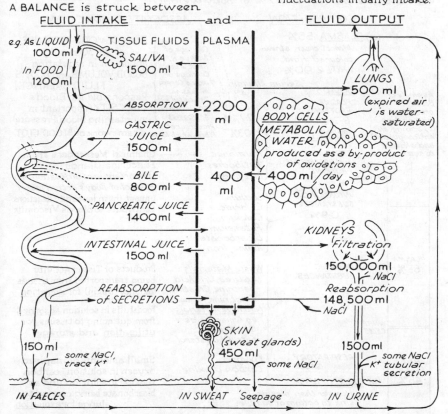

FLUID INTAKE ———— and ———— FLUID OUTPUT

e.g. As *LIQUID*
1000 ml

In *FOOD*
1200 ml

TISSUE FLUIDS   PLASMA

SALIVA
1500 ml

ABSORPTION   2200 ml

GASTRIC JUICE
1500 ml

BILE
800 ml   400 ml ← 400 ml / day

PANCREATIC JUICE
1400 ml

INTESTINAL JUICE
1500 ml

REABSORPTION
of *SECRETIONS*

150 ml

some NaCl,
trace K⁺

IN FAECES

LUNGS
500 ml
(expired air
is water-
saturated)

BODY CELLS
METABOLIC
WATER
produced as a by-product
of oxidations

KIDNEYS
Filtration
150,000 ml
NaCl
Reabsorption
148,500 ml
NaCl

1500 ml
some NaCl
K⁺ tubular
secretion

SKIN
(sweat glands)
450 ml
some NaCl

IN SWEAT   'Seepage'   *IN URINE*

*Except in Growth, Convalescence or Pregnancy, when new tissue is
being formed, an INCREASE or DECREASE in INTAKE leads to an appropriate
INCREASE or DECREASE in OUTPUT to maintain the BALANCE.*

BLOOD

Blood is the specialized fluid tissue of the *TRANSPORT SYSTEM*.
[*Specific Gravity*, 1·055 – 1·065; *pH*, 7·3 – 7·4; *Average amount*, 5 litres, *varies with body weight* (about 7·7% of body weight).

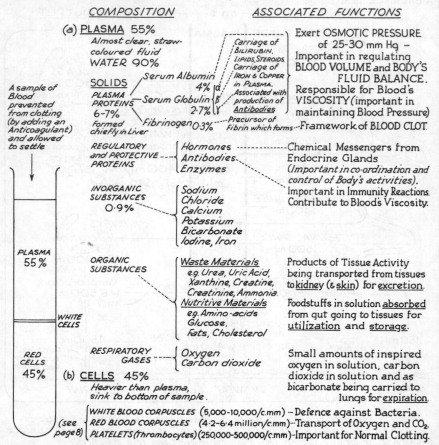

COMPOSITION

(a) <u>PLASMA</u> 55%

Almost clear, straw-coloured fluid
WATER 90%

<u>SOLIDS</u>
PLASMA
PROTEINS
6-7%

Serum Albumin 4% α
Serum Globulin 2·7% β γ
Fibrinogen 0·3%

Formed chiefly in Liver

REGULATORY and PROTECTIVE PROTEINS
{ Hormones
  Antibodies
  Enzymes

INORGANIC SUBSTANCES 0·9%
{ Sodium
  Chloride
  Calcium
  Potassium
  Bicarbonate
  Iodine, Iron

ORGANIC SUBSTANCES
<u>Waste Materials</u>
e.g. Urea, Uric Acid, Xanthine, Creatine, Creatinine, Ammonia
<u>Nutritive Materials</u>
e.g. Amino-acids Glucose, Fats, Cholesterol

RESPIRATORY GASES
{ Oxygen
  Carbon dioxide

(b) <u>CELLS</u> 45%
Heavier than plasma, sink to bottom of sample.

A sample of Blood prevented from clotting (by adding an Anticoagulant) and allowed to settle

PLASMA 55%

WHITE CELLS

RED CELLS 45%

(see page 8)

ASSOCIATED FUNCTIONS

Carriage of BILIRUBIN, LIPIDS, STEROIDS. Carriage of IRON & COPPER in PLASMA. Associated with production of Antibodies. Precursor of Fibrin which forms

Exert OSMOTIC PRESSURE of 25-30 mm Hg – Important in regulating BLOOD VOLUME and BODY'S FLUID BALANCE. Responsible for Blood's VISCOSITY (important in maintaining Blood Pressure) Framework of BLOOD CLOT.

Chemical Messengers from Endocrine Glands (*Important in co-ordination and control of Body's activities*). Important in Immunity Reactions. Contribute to Blood's Viscosity.

Products of Tissue Activity being transported from tissues to <u>kidney</u> (& <u>skin</u>) for <u>excretion</u>.

Foodstuffs in solution <u>absorbed</u> from gut going to tissues for <u>utilization</u> and <u>storage</u>.

Small amounts of inspired oxygen in solution, carbon dioxide in solution and as bicarbonate being carried to lungs for <u>expiration</u>.

WHITE BLOOD CORPUSCLES (5,000-10,000/c.mm) – Defence against Bacteria.
RED BLOOD CORPUSCLES (4·2-6·4 million/c.mm) – Transport of Oxygen and CO₂.
PLATELETS (Thrombocytes) (250,000-500,000/c.mm) – Important for Normal Clotting.

HAEMOSTASIS AND BLOOD COAGULATION

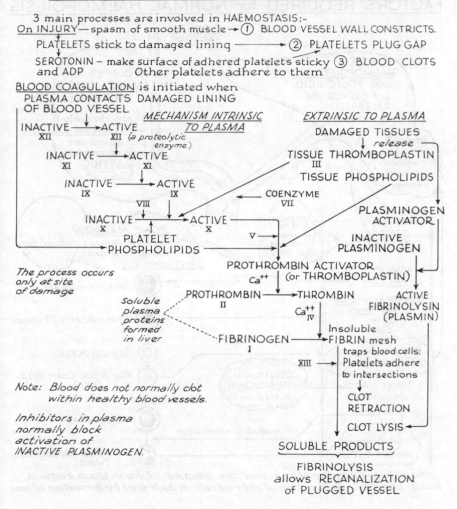

3 main processes are involved in HAEMOSTASIS:-

On INJURY— spasm of smooth muscle → (1) BLOOD VESSEL WALL CONSTRICTS.

PLATELETS stick to damaged lining ———→ (2) PLATELETS PLUG GAP

SEROTONIN – make surface of adhered platelets sticky (3) BLOOD CLOTS
and ADP   Other platelets adhere to them

BLOOD COAGULATION is initiated when
PLASMA CONTACTS DAMAGED LINING
OF BLOOD VESSEL

MECHANISM INTRINSIC
TO PLASMA

INACTIVE —→ ACTIVE
XII   XII *(a proteolytic enzyme)*

INACTIVE ——→ ACTIVE
XI   XI

INACTIVE ——→ ACTIVE
IX   IX

        VIII
INACTIVE ——→ ACTIVE
X   X

PLATELET
PHOSPHOLIPIDS

EXTRINSIC TO PLASMA

DAMAGED TISSUES
      ↓ *release*
TISSUE THROMBOPLASTIN
III
TISSUE PHOSPHOLIPIDS

←— COENZYME
VII

PLASMINOGEN
ACTIVATOR

INACTIVE
PLASMINOGEN

V →

PROTHROMBIN ACTIVATOR
Ca^{++} (or THROMBOPLASTIN)

*The process occurs
only at site
of damage*

*Soluble
plasma
proteins
formed
in liver*

PROTHROMBIN ——→ THROMBIN
II Ca^{++}_{IV}

ACTIVE
FIBRINOLYSIN
(PLASMIN)

FIBRINOGEN ——→ FIBRIN mesh
I   Insoluble

XIII —→ traps blood cells:
Platelets adhere
to intersections

CLOT
RETRACTION

CLOT LYSIS ←—

*Note: Blood does not normally clot
within healthy blood vessels.*

*Inhibitors in plasma
normally block
activation of
INACTIVE PLASMINOGEN.*

SOLUBLE PRODUCTS

FIBRINOLYSIS
allows RECANALIZATION
of PLUGGED VESSEL

61

FACTORS REQUIRED for NORMAL HAEMOPOIESIS

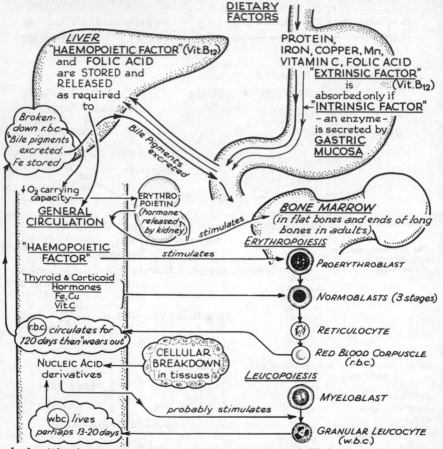

In health the number of r.b.c. and the amount of Hb in them remain fairly constant. Destruction of old red cells is balanced by formation of new.

BLOOD GROUPS

There are present in the PLASMA of some individuals substances which can cause the AGGLUTINATION *(clumping together)* and subsequent HAEMOLYSIS *(breakdown)* of the RED BLOOD CELLS of some other individuals.

If such reactions follow BLOOD TRANSFUSION the two bloods are said to be INCOMPATIBLE. Small blood vessels in lungs or brain may be blocked. Haemolysis may lead to Hb in urine and eventually to kidney failure and death.

Two FACTORS are involved in an AGGLUTINATION reaction :-
An AGGLUTINOGEN present in DONOR'S Red Blood Cell ⎱ e.g. A ⎫ or B ⎱
A specific AGGLUTININ present in RECIPIENT'S Plasma ⎰ α ⎭ or β ⎰

Obviously no such combination occurs naturally otherwise auto-agglutination would result.

In the ABO System

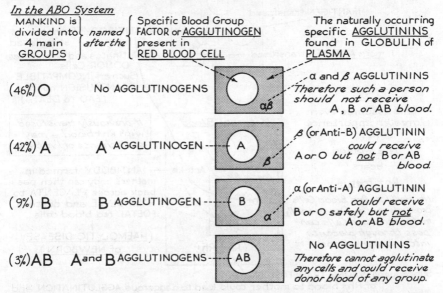

MANKIND is divided into 4 main GROUPS *named after the*	Specific Blood Group FACTOR or <u>AGGLUTINOGEN</u> present in RED BLOOD CELL	The naturally occurring specific <u>AGGLUTININS</u> found in GLOBULIN of PLASMA
(46%) O	No AGGLUTINOGENS	α and β AGGLUTININS *Therefore such a person should not receive A, B or AB blood.*
(42%) A	A AGGLUTINOGEN	β (or Anti-B) AGGLUTININ *could receive A or O but <u>not</u> B or AB blood.*
(9%) B	B AGGLUTINOGEN	α (or Anti-A) AGGLUTININ *could receive B or O safely but <u>not</u> A or AB blood.*
(3%) AB	A and B AGGLUTINOGENS	No AGGLUTININS *Therefore cannot agglutinate any cells and could receive donor blood of any group.*

<u>In Practice</u> *it is important that the DONOR's cells should not be agglutinated by the RECIPIENT's plasma. Agglutination of Recipient's cells by Donor agglutinins is less likely to occur. To avoid sub-group incompatibility Donor's blood is always matched directly with Patient's blood.*

63

RHESUS FACTOR

Over 50 different BLOOD GROUP FACTORS have been demonstrated. Not all are important in blood transfusion work.

The <u>RHESUS FACTOR</u> is important

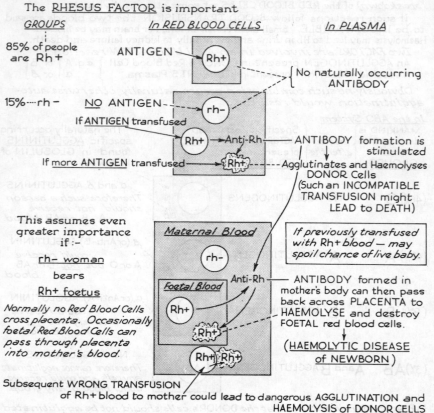

GROUPS	*In RED BLOOD CELLS*	*In PLASMA*

85% of people are Rh+

ANTIGEN — (Rh+)

{ No naturally occurring ANTIGEN

15%····rh − <u>NO ANTIGEN</u> — (rh−)

If <u>ANTIGEN</u> transfused

(Rh+) → Anti-Rh — ANTIBODY formation is stimulated

If <u>more ANTIGEN</u> transfused → (Rh+) ---- Agglutinates and Haemolyses DONOR Cells (Such an INCOMPATIBLE TRANSFUSION might LEAD TO DEATH)

This assumes even greater importance if :−

<u>rh− woman</u>

bears

<u>Rh+ foetus</u>

Normally no Red Blood Cells cross placenta. Occasionally foetal Red Blood Cells can pass through placenta into mother's blood.

Maternal Blood

(rh−)

Foetal Blood

(Rh+)

(Rh+)

(Rh+)(Rh+)

Anti-Rh

If previously transfused with Rh+ blood — may spoil chance of live baby.

ANTIBODY formed in mother's body can then pass back across PLACENTA to HAEMOLYSE and destroy FOETAL red blood cells.

(<u>HAEMOLYTIC DISEASE of NEWBORN</u>)

Subsequent WRONG TRANSFUSION of Rh+ blood to mother could lead to dangerous AGGLUTINATION and HAEMOLYSIS of DONOR CELLS within mother's own body.

There are many sub-groups in this system.

LYMPHATIC SYSTEM

All CELLS
are bathed by TISSUE FLUID.
This diffuses from CAPILLARIES.
Some returns to CAPILLARIES.
Some drains into blind-ending,
thin-walled LYMPHATICS.
It is then known as LYMPH
(similar to plasma but less protein).

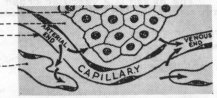

A network of LYMPHATIC VESSELS drains tissue
spaces throughout the body (except in Central Nervous
System). They unite to form LARGER and LARGER
vessels ⟶ RIGHT LYMPHATIC DUCT and LEFT
THORACIC DUCT ⟶ SUBCLAVIAN VEINS (i.e. lymph
is returned to the blood stream here)

In the course of LARGER vessels, LYMPH is
filtered through <u>LYMPH NODES</u>

AFFERENT LYMPHATICS- pour their LYMPH into
RETICULAR FRAMEWORK
of loose Sinus tissue.

MACROPHAGE cells ingest
foreign material (e.g.
carbon in lungs) or
harmful bacteria.

LYMPH NODULES produce
LYMPHOCYTES and
PLASMA CELLS

important in antibody
formation and immuno-
logical reactions

Capsule of
Fibrous Tissue

EFFERENT
LYMPHATIC — receives lymph after its
slow passage through node

MOVEMENT of LYMPH towards HEART depends partly on compression of lymphatic
vessels by muscles of limbs and partly on 'suction' created by movements of
respiration. Valves within the vessels prevent backflow. The lymphatic
tissue of the body forms an important part of the Body's Defence against
invading agents such as PROTOZOA, BACTERIA, VIRUSES, or their poisonous TOXINS.
These act as ANTIGENS stimulating ANTIBODY FORMATION — which can
subsequently destroy or neutralize the antigen.

SPLEEN

The Spleen is a vascular organ, weighing about 200 grams. It is situated in the left side of the abdomen behind the stomach and above the kidney.

FIBROUS TISSUE CAPSULE
with extensions into substance
of organ
TRABECULAE

SPLENIC ARTERY
and VEIN
and branches
TRABECULAR
ARTERY and VEIN

Near their
termination
ARTERIOLES
are surrounded
by collections of
LYMPHATIC TISSUE
("WHITE PULP")—contains
the plasma cells which
manufacture many antibodies and produces
LYMPHOCYTES→ added to blood in its
passage through spleen.

"RED PULP"— Framework of Reticular Tissue — acts
as RESERVOIR for Blood.

(Some smooth muscle in Capsule and Trabeculae may aid in expelling blood from reservoir to general circulation.)

PHAGOCYTIC CELLS of | Destroy worn out Red Blood
ELLIPSOID and other | Corpuscles, platelets and other
Reticulo-Endothelial Cells | (foreign) particles, etc.

It is difficult to trace further pathways of blood through spleen and there is much disagreement about the exact route taken. Probably Arterioles link up with VENOUS SINUSES (in the way shown).

Blood appears to percolate through the narrow spaces in the wall of the SINUS into the Red Pulp— and can get back into Sinus in same way.

During foetal life the spleen forms red and white blood cells.
It is not apparently essential to life in the adult.

CEREBROSPINAL FLUID

COMPOSITION:- Cerebrospinal Fluid (C.S.F.) is similar to blood plasma but does not clot. It is a clear watery alkaline fluid. It contains salts, glucose, some urea and creatinine, very little protein and very few lymphocytes.
VOLUME:- 120-150c.c. in man.   _SPECIFIC GRAVITY_:- 1·005-1·008.

BRAIN is covered by 4 membranes:-

DURA MATER
  Periosteal layer
  Between these is Venous Blood
  Meningeal layer
ARACHNOID MATER
  Between these is C.S.F.
PIA MATER

The _Subarachnoid Space_ - containing C.S.F. - links with Ventricles within the Brain.

ARACHNOID VILLI project into VENOUS SINUSES containing blood

C.S.F. from Foramen of Munro in Lateral Ventricle

III V.

FORMATION

×60

CHOROID PLEXUSES in Ventricles secrete C.S.F. continuously (500c.c./day).

CIRCULATION
C.S.F. formed in Lateral Vent's. joins that from IIIrd & IVth. Vent's. (see p.118) to circulate over surface of Brain and Spinal Cord in Subarachnoid space.

REABSORPTION through vascular tufts-ARACHNOID VILLI-into blood stream. [_Effective forces_:- HYDROSTATIC PRESSURE of C.S.F.- 120 mm H_2O - is greater than venous pressure in sinuses: it is aided by osmotic pull of Plasma Proteins within Plasma in returning C.S.F. to Blood Stream]

SPINAL CORD

DURA MATER
ARACHNOID MATER
PIA MATER

FUNCTIONS of C.S.F.
1. _Protective Covering_ for delicate brain tissues.
2. _Alteration of Volume_ can compensate for fluctuations in amount of blood within skull and thus keep total volume of cranial contents constant.
3. _Exchange of Metabolic substances_ between Nerve Cells and C.S.F.
                              (i.e. it receives some waste products.)

CHAPTER 4.

RESPIRATORY SYSTEM

All living cells require to get OXYGEN from the fluid around them and to get rid of CARBON DIOXIDE to it.

<u>INTERNAL RESPIRATION</u> is the exchange of these gases between tissue cells and their fluid environment.

<u>EXTERNAL RESPIRATION</u> is the exchange of these gases *(oxygen and carbon dioxide)* between the body and the external environment.

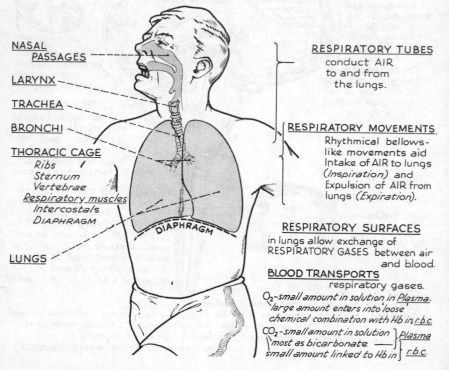

NASAL PASSAGES

LARYNX

TRACHEA

BRONCHI

THORACIC CAGE
  Ribs
  Sternum
  Vertebrae
  <u>*Respiratory muscles*</u>
  Intercostals
  DIAPHRAGM

LUNGS

DIAPHRAGM

RESPIRATORY TUBES
conduct AIR
to and from
the lungs.

RESPIRATORY MOVEMENTS
Rhythmical bellows-like movements aid
Intake of AIR to lungs
(Inspiration) and
Expulsion of AIR from
lungs *(Expiration)*.

RESPIRATORY SURFACES
in lungs allow exchange of
RESPIRATORY GASES between air
and blood.

BLOOD TRANSPORTS
respiratory gases.
O_2-*small amount in solution in* <u>*Plasma*</u>.
*large amount enters into 'loose'
chemical combination with Hb in* <u>*r.b.c.*</u>
CO_2-*small amount in solution* } <u>*Plasma*</u>
most as bicarbonate ——— }
small amount linked to Hb in } <u>*r.b.c.*</u>

AIR CONDUCTING PASSAGES

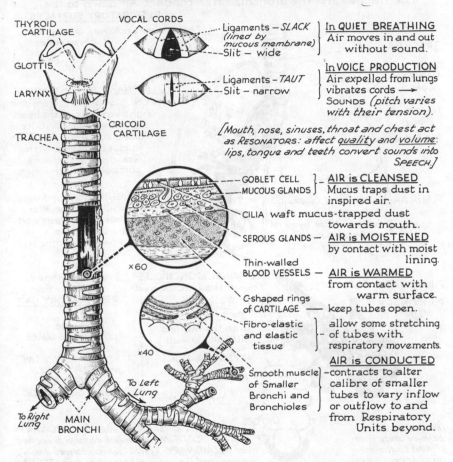

VOCAL CORDS

THYROID CARTILAGE

GLOTTIS

LARYNX

CRICOID CARTILAGE

TRACHEA

Ligaments – *SLACK* (lined by mucous membrane)
Slit – wide

In **QUIET BREATHING** Air moves in and out without sound.

Ligaments – *TAUT*
Slit – narrow

In **VOICE PRODUCTION** Air expelled from lungs vibrates cords ⟶ Sounds (*pitch varies with their tension*).

[Mouth, nose, sinuses, throat and chest act as RESONATORS: affect quality and volume: lips, tongue and teeth convert sounds into SPEECH.]

GOBLET CELL
MUCOUS GLANDS

AIR is CLEANSED Mucus traps dust in inspired air.

CILIA waft mucus-trapped dust towards mouth.

SEROUS GLANDS —

AIR is MOISTENED by contact with moist lining.

Thin-walled BLOOD VESSELS —

AIR is WARMED from contact with warm surface.

×60

C-shaped rings of CARTILAGE — keep tubes open.

Fibro-elastic and elastic tissue

allow some stretching of tubes with respiratory movements.

×40

AIR is CONDUCTED — contracts to alter calibre of smaller tubes to vary inflow or outflow to and from Respiratory Units beyond.

Smooth muscle of Smaller Bronchi and Bronchioles

To *Left* Lung

To *Right* Lung   MAIN BRONCHI

LUNGS: RESPIRATORY SURFACES

The Trachea and the Bronchial 'Tree' conduct Air down to the
RESPIRATORY SURFACES.

There is no exchange of gases in these tubes.

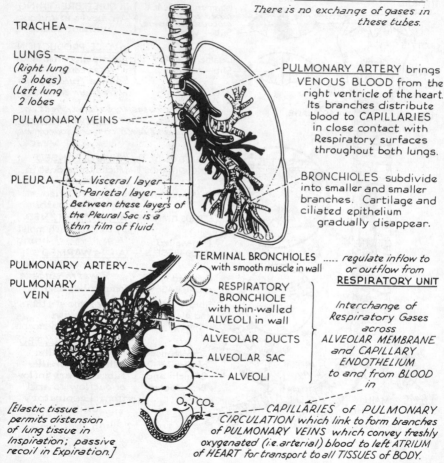

TRACHEA

LUNGS
(Right lung
3 lobes)
(Left lung
2 lobes)

PULMONARY VEINS

PLEURA — Visceral layer —
Parietal layer —
*Between these layers of
the Pleural Sac is a
thin film of fluid.*

PULMONARY ARTERY brings
VENOUS BLOOD from the
right ventricle of the heart.
Its branches distribute
blood to CAPILLARIES
in close contact with
Respiratory surfaces
throughout both lungs.

BRONCHIOLES subdivide
into smaller and smaller
branches. Cartilage and
ciliated epithelium
gradually disappear.

PULMONARY ARTERY

PULMONARY
VEIN

TERMINAL BRONCHIOLES
with smooth muscle in wall

..... *regulate inflow to
or outflow from*
RESPIRATORY UNIT

RESPIRATORY
BRONCHIOLE
with thin-walled
ALVEOLI in wall

ALVEOLAR DUCTS

ALVEOLAR SAC

ALVEOLI

*Interchange of
Respiratory Gases
across
ALVEOLAR MEMBRANE
and CAPILLARY
ENDOTHELIUM
to and from BLOOD
in*

[*Elastic tissue
permits distension
of lung tissue in
Inspiration; passive
recoil in Expiration.*]

O_2 ↑ CO_2

CAPILLARIES *of* PULMONARY
CIRCULATION *which link to form branches
of* PULMONARY VEINS *which convey freshly
oxygenated (i.e. arterial) blood to left* ATRIUM
of HEART *for transport to all* TISSUES *of BODY.*

THORAX

The Thorax (*or* chest) is the closed cavity which contains the LUNGS, HEART and Great Vessels.

It is enclosed and bounded
<u>ABOVE</u> by the upper RIBS and tissues of the neck;
<u>AT THE SIDES</u> by the RIBS and INTERCOSTAL MUSCLES;
<u>AT THE BACK</u> by the RIBS and VERTEBRAL COLUMN (or *back bone*);
<u>IN FRONT</u> by the RIBS, COSTAL CARTILAGES and STERNUM (*or breast bone*);
<u>BELOW</u> by the DIAPHRAGM (a strong dome-shaped sheet of skeletal muscle which separates the thoracic cavity from the abdominal cavity).

Parietal

Visceral

Crura

It is lined by a thin moist membrane — the PLEURA — the inner layer of which invests the LUNGS. In health there is a thin film of fluid between these two pleural layers.

Capillaries of the <u>pulmonary</u> and <u>not the systemic circulation</u> supply the visceral layer. Because their blood pressure (5-10mm Hg) is less than the osmotic 'pull' of the plasma proteins (25 mm Hg) fluid is continually 'suctioned' into the capillaries from the INTRAPLEURAL SPACE. This leads to the subatmospheric or <u>negative intrapleural pressure</u> essential for normal breathing.

DIMENSIONS of thoracic cage and the PRESSURE between pleural surfaces change rhythmically about 18-20 times a minute with the MOVEMENTS of RESPIRATION —— AIR MOVEMENT in and out of the lungs follows
                                                            passively.

MECHANISM of BREATHING

The rhythmical changes in the CAPACITY of the Thorax are brought about by MUSCULAR ACTION. The changes in LUNG VOLUME with INTAKE or EXPULSION of air follow passively.

In NORMAL QUIET BREATHING

INSPIRATION

EXTERNAL INTERCOSTAL MUSCLES actively contract—
– ribs and sternum move upwards and outwards
– width of chest increases from side to side and from front to back.

DIAPHRAGM contracts—
– descends
– depth of chest increases.

CAPACITY of THORAX is INCREASED

↓

PRESSURE between PLEURAL SURFACES *(already negative)* is REDUCED from –2 to –6 mm Hg *(i.e. an increased "suction pull" is exerted on LUNG TISSUE)*

↓

ELASTIC TISSUE of LUNGS is STRETCHED

↓

LUNGS EXPAND to fill THORACIC CAVITY

↓

AIR PRESSURE within ALVEOLI is now less than atmospheric pressure

↓

AIR is sucked into ALVEOLI from ATMOSPHERE

EXPIRATION

EXTERNAL INTERCOSTAL MUSCLES relax—
– ribs and sternum move downwards and inwards
– width of chest diminishes.

DIAPHRAGM relaxes—
– ascends
– depth of chest diminishes.

CAPACITY of THORAX is DECREASED

↓

PRESSURE between PLEURAL SURFACES is INCREASED from –6 to –2 mm Hg *(i.e. less pull is exerted on LUNG TISSUE)*

↓

ELASTIC TISSUE of LUNGS RECOILS

↓

AIR PRESSURE within ALVEOLI is now greater than atmospheric pressure

↓

AIR is forced out of ALVEOLI to ATMOSPHERE

In FORCED BREATHING

Muscles of Nostrils and round Glottis may contract to aid entrance of air to lungs.
Extensors of Vertebral Column may aid inspiration.
Muscles of Neck contract—
– move 1ST rib upwards
(and sternum upwards and forwards)

Internal Intercostals may contract—
– move ribs downwards more actively.
Abdominal muscles contract—
– actively aid ascent of diaphragm.

CAPACITY of LUNGS

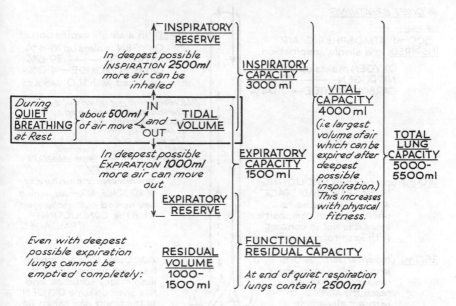

INSPIRATORY RESERVE — In deepest possible *INSPIRATION 2500ml* more air can be inhaled

INSPIRATORY CAPACITY 3000 ml

During **QUIET BREATHING** *at Rest* — *about 500ml of air move* IN *and* OUT — **TIDAL VOLUME**

VITAL CAPACITY 4000 ml
(i.e largest volume of air which can be expired after deepest possible inspiration.) This increases with physical fitness.

EXPIRATORY RESERVE — In deepest possible *EXPIRATION 1000ml* more air can move out

EXPIRATORY CAPACITY 1500 ml

TOTAL LUNG CAPACITY 5000-5500ml

Even with deepest possible expiration lungs cannot be emptied completely:

RESIDUAL VOLUME 1000-1500 ml

FUNCTIONAL RESIDUAL CAPACITY — At end of quiet respiration lungs contain 2500ml

*Rate and depth of breathing vary with circumstance —
e.g. amount of air moving in and out increases with muscular
activity and both inspiratory and expiratory reserves become
smaller.*

*At rest a normal male adult breathes in and out say 16 times per
minute. The amount of air breathed in per minute is therefore 500ml×16,
i.e. 8000ml or 8 litres — This is the RESPIRATORY MINUTE VOLUME or
PULMONARY VENTILATION. In exercise it may go up to as much as
200 litres. These values are about 25% lower in women.*

COMPOSITION of RESPIRED AIR

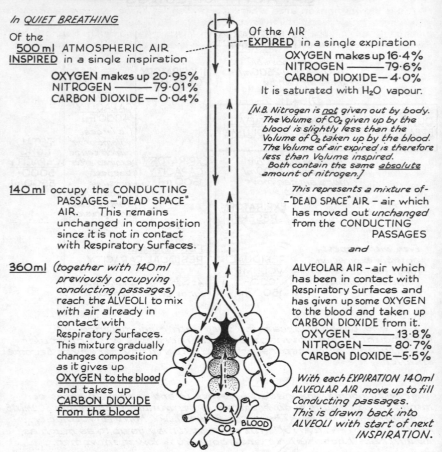

In _QUIET BREATHING_

Of the
500 ml ATMOSPHERIC AIR
INSPIRED in a single inspiration

    OXYGEN makes up 20·95%
    NITROGEN ———— 79·01%
    CARBON DIOXIDE— 0·04%

Of the AIR
EXPIRED in a single expiration
    OXYGEN makes up 16·4%
    NITROGEN ————79·6%
    CARBON DIOXIDE— 4·0%
It is saturated with H_2O vapour.

*[N.B. Nitrogen is <u>not</u> given out by body.
The Volume of CO_2 given up by the
blood is slightly less than the
Volume of O_2 taken up by the blood.
The Volume of air expired is therefore
less than Volume inspired.
  Both contain the same <u>absolute</u>
amount of nitrogen.]*

140 ml occupy the CONDUCTING
    PASSAGES – "DEAD SPACE"
    AIR.   This remains
    unchanged in composition
    since it is not in contact
    with Respiratory Surfaces.

This represents a mixture of-
-"DEAD SPACE" AIR – air which
has moved out *unchanged*
from the CONDUCTING
        PASSAGES

and

360 ml *(together with 140 ml
previously occupying
conducting passages)*
reach the ALVEOLI to mix
with air already in
contact with
Respiratory Surfaces.
This mixture gradually
changes composition
as it gives up
<u>OXYGEN to the blood</u>
and takes up
<u>CARBON DIOXIDE</u>
<u>from the blood</u>

ALVEOLAR AIR – air which
has been in contact with
Respiratory Surfaces and
has given up some OXYGEN
to the blood and taken up
CARBON DIOXIDE from it.
    OXYGEN ———— 13·8%
    NITROGEN———— 80·7%
    CARBON DIOXIDE—5·5%

*With each EXPIRATION 140ml
ALVEOLAR AIR move up to fill
Conducting passages.
This is drawn back into
ALVEOLI with start of next
      INSPIRATION.*

In <u>_DEEP BREATHING_</u> *the composition of Air breathed out with each
          expiration comes closer to that of Alveolar Air.*

INTERCHANGE of RESPIRATORY GASES

A gas moves from an area where it is present at *higher concentration* to an area where it is present at *lower concentration*. The movement of gas molecules continues till the pressure exerted by them is the same throughout both areas. DRY Atmospheric Air (*at sea level*) has a pressure of 760 mm Hg.

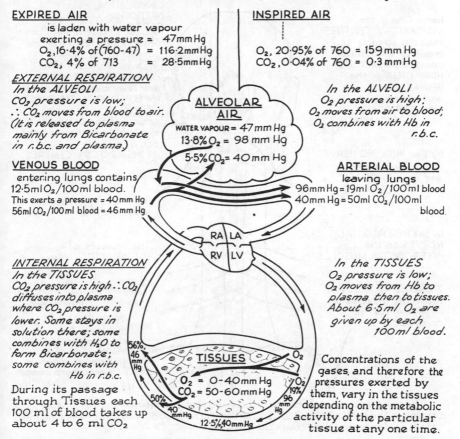

EXPIRED AIR
is laden with water vapour
exerting a pressure = 47 mm Hg
O_2, 16·4% of (760 - 47) = 116·2 mm Hg
CO_2, 4% of 713 = 28·5 mm Hg

EXTERNAL RESPIRATION
In the ALVEOLI
CO_2 pressure is low;
∴ CO_2 moves from blood to air.
(It is released to plasma
mainly from Bicarbonate
in r.b.c. and plasma)

VENOUS BLOOD
entering lungs contains
12·5 ml O_2/100 ml blood.
This exerts a pressure = 40 mm Hg
56 ml CO_2/100 ml blood = 46 mm Hg

INSPIRED AIR

O_2, 20·95% of 760 = 159 mm Hg
CO_2, 0·04% of 760 = 0·3 mm Hg

In the ALVEOLI
O_2 pressure is high;
O_2 moves from air to blood;
O_2 combines with Hb in
r.b.c.

ALVEOLAR AIR
WATER VAPOUR = 47 mm Hg
13·8% O_2 = 98 mm Hg
5·5% CO_2 = 40 mm Hg

ARTERIAL BLOOD
leaving lungs
96 mm Hg = 19 ml O_2/100 ml blood
40 mm Hg = 50 ml CO_2/100 ml blood

RA LA
RV LV

INTERNAL RESPIRATION
In the TISSUES
CO_2 pressure is high ∴ CO_2
diffuses into plasma
where CO_2 pressure is
lower. Some stays in
solution there; some
combines with H_2O to
form Bicarbonate;
some combines with
Hb in r.b.c.

During its passage
through Tissues each
100 ml of blood takes up
about 4 to 6 ml CO_2

TISSUES
O_2 = 0-40 mm Hg
CO_2 = 50-60 mm Hg

56% 46 mm Hg
50%
40 mm Hg
12·5%,40 mm Hg
7 O_2 19% 96 mm Hg

In the TISSUES
O_2 pressure is low;
O_2 moves from Hb to
plasma then to tissues.
About 6·5 ml O_2 are
given up by each
100 ml blood.

Concentrations of the
gases, and therefore the
pressures exerted by
them, vary in the tissues
depending on the metabolic
activity of the particular
tissue at any one time.

NERVOUS CONTROL of RESPIRATORY MOVEMENTS

Normal Respiratory Movements are Involuntary. They are carried out automatically (i.e. without conscious control) through the rhythmical discharge of nerve impulses from <u>CONTROLLING CENTRES</u> in the <u>BRAIN</u>.

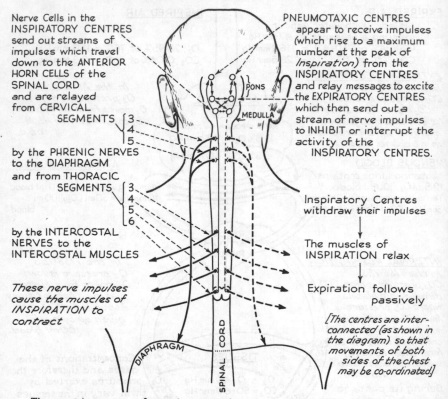

Nerve Cells in the INSPIRATORY CENTRES send out streams of impulses which travel down to the ANTERIOR HORN CELLS of the SPINAL CORD and are relayed from CERVICAL SEGMENTS {3 4 5

by the PHRENIC NERVES to the DIAPHRAGM and from THORACIC SEGMENTS {3 4 5 6

by the INTERCOSTAL NERVES to the INTERCOSTAL MUSCLES

These nerve impulses cause the muscles of INSPIRATION to contract

PNEUMOTAXIC CENTRES appear to receive impulses (which rise to a maximum number at the peak of *Inspiration*) from the INSPIRATORY CENTRES and relay messages to excite the EXPIRATORY CENTRES which then send out a stream of nerve impulses to INHIBIT or interrupt the activity of the INSPIRATORY CENTRES.

PONS

MEDULLA

Inspiratory Centres withdraw their impulses

↓

The muscles of INSPIRATION relax

↓

Expiration follows passively

[The centres are interconnected (as shown in the diagram) so that movements of both sides of the chest may be co-ordinated]

DIAPHRAGM

SPINAL CORD

The most important factor in regulating the activity of the respiratory centres is the level of CARBON DIOXIDE in the blood — an increase stimulates; a decrease depresses the centre.

VOLUNTARY and REFLEX FACTORS in the REGULATION of RESPIRATION

Although fundamentally automatic and regulated by chemical factors in the blood, ingoing impulses from many parts of the body also modify the activity of the RESPIRATORY CENTRES and consequently alter the outgoing impulses to the Respiratory muscles to co-ordinate RHYTHM, RATE or DEPTH of breathing with other activities of the body.

Impulses from **HIGHER CENTRES** - PSYCHIC and EMOTIONAL INFLUENCES
{
Voluntary alterations in Breathing.
Interruptions of expiration in SPEECH and SINGING.
Deep inspiration then short spasmodic expirations in LAUGHTER and WEEPING.
Prolonged expiration in SIGHING.
Deep inspiration with mouth open in YAWNING.
Slow shallow breathing in SUSPENSE and CONCENTRATION.
Rapid breathing in FEAR and EXCITEMENT.
}

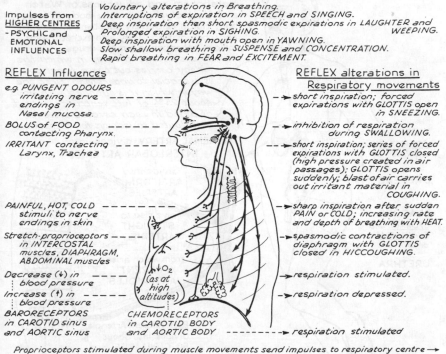

REFLEX Influences

e.g. *PUNGENT ODOURS*
    irritating nerve ──────
    endings in
    Nasal mucosa.

BOLUS of FOOD ──────
    contacting Pharynx.

IRRITANT contacting ──────
    Larynx, Trachea

PAINFUL, HOT, COLD ────────
    stimuli to nerve
    endings in skin

Stretch-proprioceptors ────
    in INTERCOSTAL
    muscles, DIAPHRAGM,
    ABDOMINAL muscles

Decrease (↓) in ────
    blood pressure

Increase (↑) in ─────
    blood pressure

BARORECEPTORS in CAROTID sinus and AORTIC sinus

↓O₂ (as at high altitudes)

CHEMORECEPTORS in CAROTID BODY and AORTIC BODY

REFLEX alterations in Respiratory movements

─→ short inspiration; forced expirations with GLOTTIS open in SNEEZING.

─→ inhibition of respiration during SWALLOWING.

──→ short inspiration; series of forced expirations with GLOTTIS closed (high pressure created in air passages); GLOTTIS opens suddenly; blast of air carries out irritant material in COUGHING.

─→ sharp inspiration after sudden PAIN or COLD; increasing rate and depth of breathing with HEAT.

─→ spasmodic contractions of diaphragm with GLOTTIS closed in HICCOUGHING.

─→ respiration stimulated.

─→ respiration depressed.

────→ respiration stimulated

Proprioceptors stimulated during muscle movements send impulses to respiratory centre → ↑rate and depth of breathing. (N.B. This occurs with active or passive movements of limbs.)

In normal breathing respiratory rate and rhythm are thought to be influenced rhythmically by the Hering-Breuer Reflex.

| Distension of alveoli at end of inspiration | stimulates stretch receptors in bronchioles | Stream of ingoing impulses passes along vagus nerves to depress inspiratory centres | Withdrawal of outgoing impulses to respiratory muscles → expiration |

CHAPTER 5.
EXCRETORY SYSTEM

Together with the *Respiratory System* and the *Skin*, the KIDNEYS are the chief *Excretory organs* of the body.

The Kidneys excrete waste products of metabolism and adjust loss of Water and Electrolytes from the body to keep body fluids relatively constant in amount and composition. They also excrete some toxic substances, e.g. many drugs.

To understand the way in which the kidney carries out these functions, it is essential to understand first the way in which it is supplied with blood. About 25% of left ventricle's output of blood in each cardiac cycle is distributed to kidneys for filtration.

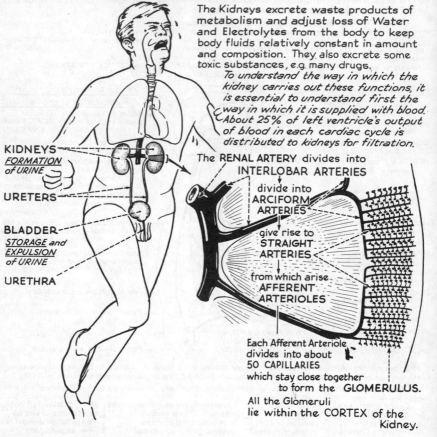

KIDNEYS
FORMATION
of URINE

URETERS

BLADDER
STORAGE and
EXPULSION
of URINE

URETHRA

The **RENAL ARTERY** divides into
INTERLOBAR ARTERIES

divide into
ARCIFORM ARTERIES

give rise to
STRAIGHT ARTERIES

from which arise
AFFERENT ARTERIOLES

Each Afferent Arteriole divides into about 50 CAPILLARIES which stay close together to form the **GLOMERULUS**.

All the Glomeruli lie within the CORTEX of the Kidney.

KIDNEY

Each Kidney contains approximately one million microscopic units — *NEPHRONS* – which form URINE.

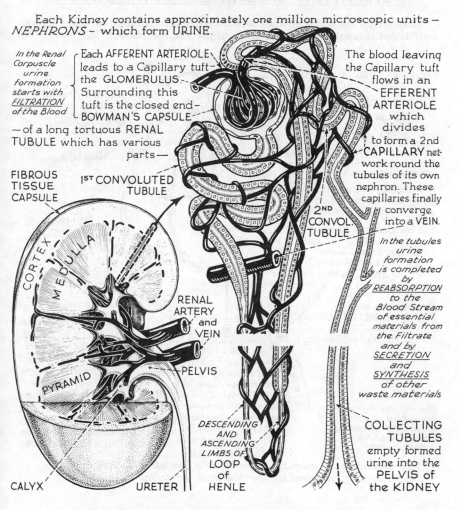

In the Renal Corpuscle urine formation starts with FILTRATION of the Blood

Each AFFERENT ARTERIOLE leads to a Capillary tuft – the GLOMERULUS. Surrounding this tuft is the closed end – BOWMAN'S CAPSULE

—of a long tortuous RENAL TUBULE which has various parts—

The blood leaving the Capillary tuft flows in an EFFERENT ARTERIOLE which divides

to form a 2nd CAPILLARY network round the tubules of its own nephron. These capillaries finally converge into a VEIN.

FIBROUS TISSUE CAPSULE

1ST CONVOLUTED TUBULE

CORTEX

MEDULLA

2ND CONVOL. TUBULE

In the tubules urine formation is completed by REABSORPTION to the Blood Stream of essential materials from the Filtrate and by SECRETION and SYNTHESIS of other waste materials

RENAL ARTERY and VEIN

PELVIS

PYRAMID

DESCENDING AND ASCENDING LIMBS OF LOOP of HENLE

COLLECTING TUBULES empty formed urine into the PELVIS of the KIDNEY

CALYX

URETER

FORMATION of URINE — 1. FILTRATION

About 25% of the left ventricle's total output of blood in each cardiac cycle is distributed through the Renal Arteries to the Kidneys for FILTRATION.

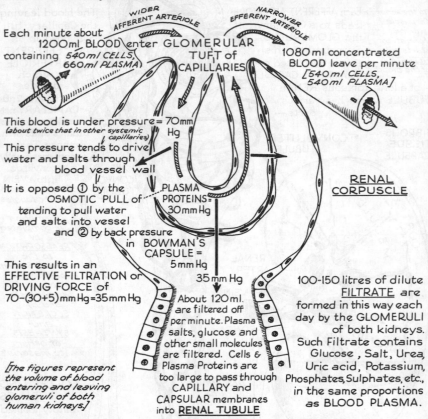

WIDER AFFERENT ARTERIOLE

NARROWER EFFERENT ARTERIOLE

Each minute about **1200ml BLOOD** enter GLOMERULAR TUFT of CAPILLARIES containing *540ml CELLS, 660ml PLASMA*

1080 ml concentrated BLOOD leave per minute [540ml CELLS, 540ml PLASMA]

This blood is under pressure = 70mm Hg *(about twice that in other systemic capillaries)*

This pressure tends to drive water and salts through blood vessel wall

It is opposed ① by the OSMOTIC PULL of PLASMA PROTEINS = 30mm Hg tending to pull water and salts into vessel and ② by back pressure in BOWMAN'S CAPSULE = 5mm Hg

35 mm Hg

RENAL CORPUSCLE

This results in an EFFECTIVE FILTRATION or DRIVING FORCE of 70−(30+5)mm Hg = 35mm Hg

About 120 ml. are filtered off per minute. Plasma salts, glucose and other small molecules are filtered. Cells & Plasma Proteins are too large to pass through CAPILLARY and CAPSULAR membranes into **RENAL TUBULE**

[The figures represent the volume of blood entering and leaving glomeruli of both human kidneys.]

100-150 litres of dilute <u>FILTRATE</u> are formed in this way each day by the GLOMERULI of both kidneys. Such Filtrate contains Glucose, Salt, Urea, Uric acid, Potassium, Phosphates, Sulphates, etc., in the same proportions as BLOOD PLASMA.

The Glomerular membrane acts as a simple FILTER — i.e. no energy is used up by the cells in filtration.

FORMATION of URINE – 2. CONCENTRATION

As it passes along the TUBULE the FILTRATE is CONCENTRATED and essential substances are CONSERVED. The Tubular Epithelium reabsorbs water and selected materials into the blood stream. Much of this activity is regulated by hormones *(see pp. 90-93,98,99)*. By varying the amount reabsorbed the composition of urine is adjusted to meet the body's needs at any one moment.

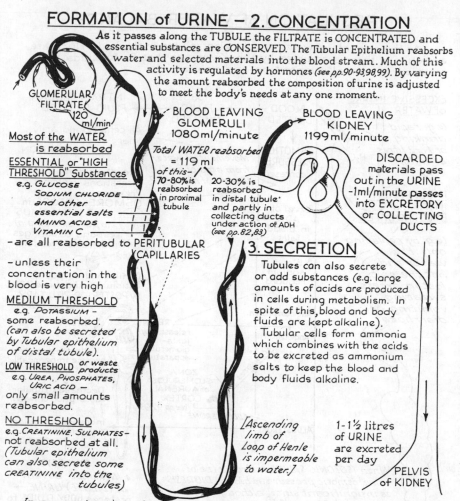

GLOMERULAR FILTRATE
120 ml/min

BLOOD LEAVING GLOMERULI
1080 ml/minute

BLOOD LEAVING KIDNEY
1199 ml/minute

Total WATER reabsorbed = 119 ml

of this – 70-80% is reabsorbed in proximal tubule

20-30% is reabsorbed in distal tubule and partly in collecting ducts under action of ADH *(see pp. 82,83)*

DISCARDED materials pass out in the URINE -1ml/minute passes into EXCRETORY or COLLECTING DUCTS

Most of the WATER is reabsorbed

ESSENTIAL or "HIGH THRESHOLD" Substances
e.g. *GLUCOSE*
SODIUM CHLORIDE and other essential salts
AMINO ACIDS
VITAMIN C

- are all reabsorbed to PERITUBULAR CAPILLARIES

- unless their concentration in the blood is very high

MEDIUM THRESHOLD
e.g. *POTASSIUM* - some reabsorbed. *(can also be secreted by Tubular epithelium of distal tubule).*

LOW THRESHOLD or waste products
e.g. *UREA, PHOSPHATES, URIC ACID* - only small amounts reabsorbed.

NO THRESHOLD
e.g. *CREATININE. SULPHATES* - not reabsorbed at all. *(Tubular epithelium can also secrete some CREATININE into the tubules)*

3. SECRETION

Tubules can also secrete or add substances (e.g. large amounts of acids are produced in cells during metabolism. In spite of this, blood and body fluids are kept alkaline).

Tubular cells form ammonia which combines with the acids to be excreted as ammonium salts to keep the blood and body fluids alkaline.

[Ascending limb of Loop of Henle is impermeable to water.]

1-1½ litres of URINE are excreted per day

PELVIS of KIDNEY

In concentrating the filtrate and in secretion of certain substances, the tubular epithelial cells use energy, i.e. they do work.

81

REGULATION of WATER BALANCE

As the amounts of water and/or electrolytes in the body fluctuate, excretion of them is adjusted by the *Kidney* so that Body Fluids are restored to normal composition and volume.

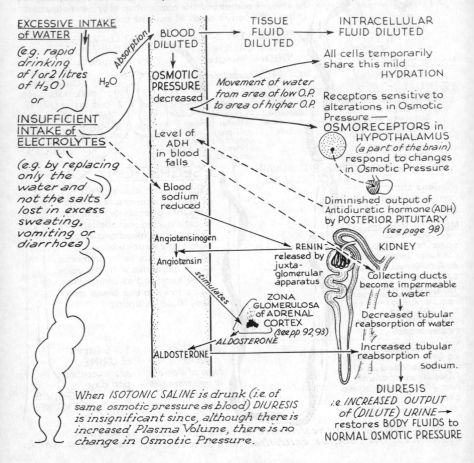

EXCESSIVE INTAKE of WATER
(e.g. rapid drinking of 1 or 2 litres of H_2O)

or

INSUFFICIENT INTAKE of ELECTROLYTES

(e.g. by replacing only the water and not the salts lost in excess sweating, vomiting or diarrhoea)

Absorption

H_2O

BLOOD DILUTED → TISSUE FLUID DILUTED → INTRACELLULAR FLUID DILUTED

All cells temporarily share this mild HYDRATION

OSMOTIC PRESSURE decreased

Movement of water from area of low O.P. to area of higher O.P.

Receptors sensitive to alterations in Osmotic Pressure — OSMORECEPTORS in HYPOTHALAMUS *(a part of the brain)* respond to changes in Osmotic Pressure.

Level of ADH in blood falls

Blood sodium reduced

Diminished output of Antidiuretic hormone (ADH) by POSTERIOR PITUITARY *(see page 98)*

Angiotensinogen

Angiotensin

RENIN released by juxta-glomerular apparatus

KIDNEY

Collecting ducts become impermeable to water

stimulates

ZONA GLOMERULOSA of ADRENAL CORTEX *(see pp 92, 93)*

ALDOSTERONE

Decreased tubular reabsorption of water

Increased tubular reabsorption of sodium.

DIURESIS *i.e. INCREASED OUTPUT of (DILUTE) URINE →* restores BODY FLUIDS to NORMAL OSMOTIC PRESSURE

When ISOTONIC SALINE is drunk (i.e. of same osmotic pressure as blood) DIURESIS is insignificant since, although there is increased Plasma Volume, there is no change in Osmotic Pressure.

82

REGULATION of WATER BALANCE

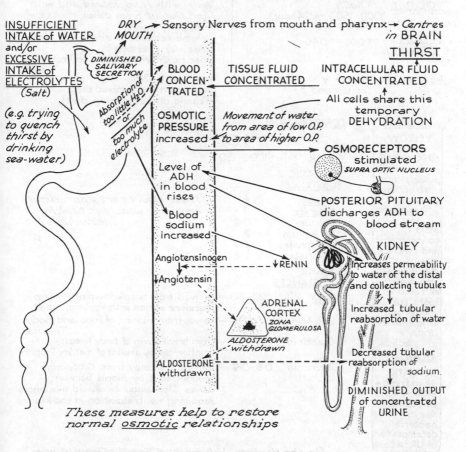

INSUFFICIENT INTAKE of WATER and/or EXCESSIVE INTAKE of ELECTROLYTES (Salt)

(e.g. trying to quench thirst by drinking sea-water)

DRY MOUTH → Sensory Nerves from mouth and pharynx → Centres in BRAIN

THIRST

DIMINISHED SALIVARY SECRETION

Absorption of too little H₂O or too much electrolyte

BLOOD CONCENTRATED

TISSUE FLUID CONCENTRATED

INTRACELLULAR FLUID CONCENTRATED

All cells share this temporary DEHYDRATION

OSMOTIC PRESSURE increased

Movement of water from area of low O.P. to area of higher O.P.

OSMORECEPTORS stimulated SUPRA OPTIC NUCLEUS

Level of ADH in blood rises

POSTERIOR PITUITARY discharges ADH to blood stream

Blood sodium increased

KIDNEY Increases permeability to water of the distal and collecting tubules

Angiotensinogen

↓RENIN

↓Angiotensin

Increased tubular reabsorption of water

ADRENAL CORTEX ZONA GLOMERULOSA

ALDOSTERONE withdrawn

Decreased tubular reabsorption of sodium.

ALDOSTERONE withdrawn

DIMINISHED OUTPUT of concentrated URINE

These measures help to restore normal osmotic relationships

These measures serve to restore Blood Volume; and to restore Body Fluids to normal OSMOTIC PRESSURE.

URINE

VOLUME: *In Adult*
1000 - 1500 ml/24 hours

SPECIFIC GRAVITY   1·001 - 1·040

Vary with *Fluid Intake* and with *Fluid Output* from other routes — Skin, Lungs, Gut.
[Volume reduced during *Sleep* and *Muscular Exercise* :
Specific Gravity greater on Protein diet.]

REACTION   Normally slightly acid — (*pH around 6*)

Varies with *Diet*
[acid on ordinary mixed diet:
alkaline on vegetarian diet].

COLOUR

YELLOW due to UROCHROME pigment-probably from destruction of tissue protein.

More concentrated and DARKER in early morning — less water excreted at night but unchanged amounts of Urinary Solids.

ODOUR

AROMATIC when fresh→ AMMONIACAL on standing due to bacterial decomposition of UREA to AMMONIA.

COMPOSITION

Grams excreted in 24 hours

WATER 96%   1000–1500

INORGANIC SUBSTANCES

Sodium	6
Chloride	7
Calcium	0·2
Potassium	2
Phosphates	1·7
Sulphates	1·8

[These figures are approximate and vary widely in healthy individuals]

ORGANIC SUBSTANCES

Urea   2%   20-30 — derived from breakdown of *Protein* — therefore varies with Protein in Diet.

Uric Acid   0·6 — comes from *Purine* of Food and Body Tissues.

Creatinine   1·2 — from breakdown of Body Tissues; uninfluenced by amount of dietary Protein.

Ammonia   0·5-0·9 — formed in Kidney from *Glutamine* brought to it by Blood Stream ; varies with amounts of acid substances requiring neutralization in the Kidney.

[In the Newborn, Volume and Specific Gravity are low and Composition varies.]

URINARY BLADDER and URETERS

A resistant, distensible *Transitional Epithelium* lines all *Urinary Passages.*

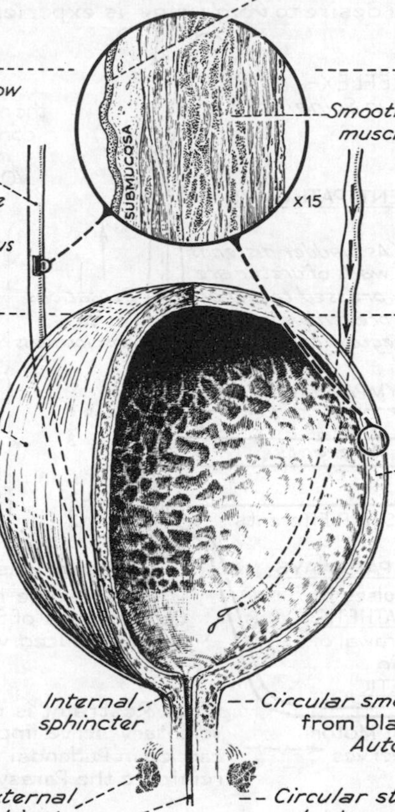

URETERS-------
Long, narrow muscular tubes with Outer Fibrous Tissue Coat and Inner Mucous Membrane

------- convey URINE from KIDNEYS to BLADDER.

Smooth muscle coats– Slow waves of contraction *(every 10 seconds)* propel urine along ureter. *1-5 small 'spurts' enter bladder per minute.*

SUBMUCOSA

x15

BLADDER----
Hollow muscular organ

---- acts as RESERVOIR for URINE. *[Size and position vary with amount of urine stored (120 - 320 c.c.).]*

Smooth muscle coats– <u>distend</u> as urine collects; <u>contract</u> periodically to expel urine to urethra.

Internal sphincter

-- Circular smooth muscle — guards exit from bladder *(under control of Autonomic Nervous System)*

External sphincter

-- Circular striated muscle *(under voluntary control - Central Nervous System)*

URETHRA-------------
Membranous tube

---------- conveys URINE to EXTERIOR.

STORAGE and EXPULSION of URINE

URINE is formed continuously by the KIDNEYS.   It collects, drop by drop, in the URINARY BLADDER which expands to hold about 300 ml. When the Bladder is full the desire to *void urine* is experienced.

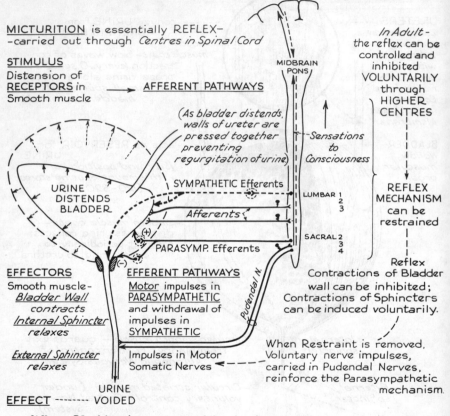

MICTURITION is essentially REFLEX— —carried out through *Centres in Spinal Cord*

In Adult — the reflex can be controlled and inhibited VOLUNTARILY through HIGHER CENTRES

STIMULUS Distension of **RECEPTORS** in Smooth muscle → AFFERENT PATHWAYS

MIDBRAIN PONS

(As bladder distends, walls of ureter are pressed together preventing regurgitation of urine)

—*Sensations to Consciousness*

URINE DISTENDS BLADDER

SYMPATHETIC Efferents

LUMBAR 1 2 3

Afferents

REFLEX MECHANISM can be restrained

PARASYMP. Efferents

SACRAL 2 3 4

EFFECTORS Smooth muscle— *Bladder Wall contracts* *Internal Sphincter relaxes*

EFFERENT PATHWAYS Motor impulses in PARASYMPATHETIC and withdrawal of impulses in SYMPATHETIC

Reflex Contractions of Bladder wall can be inhibited; Contractions of Sphincters can be induced voluntarily.

External Sphincter relaxes

Pudendal N.

Impulses in Motor Somatic Nerves ←

When Restraint is removed, Voluntary nerve impulses, carried in Pudendal Nerves, reinforce the Parasympathetic mechanism.

URINE **EFFECT** ------- VOIDED

When Bladder is empty and beginning to fill — *Inhibition of Parasympathetic* } *Relaxation of Bladder Wall.* *Stimulation of Sympathetic* } *Constriction of Sphincters.*

CHAPTER 6

ENDOCRINE SYSTEM

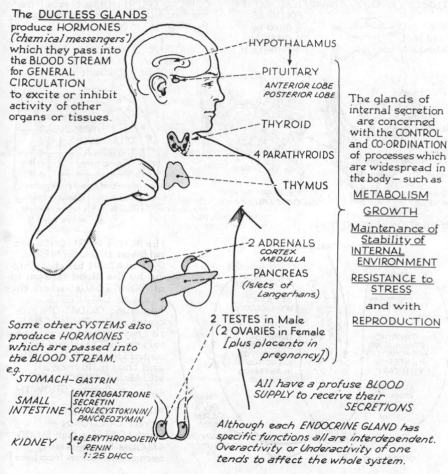

The <u>DUCTLESS GLANDS</u> produce HORMONES ("*chemical messengers*") which they pass into the BLOOD STREAM for GENERAL CIRCULATION to excite or inhibit activity of other organs or tissues.

HYPOTHALAMUS

PITUITARY
ANTERIOR LOBE
POSTERIOR LOBE

THYROID

4 PARATHYROIDS

THYMUS

2 ADRENALS
CORTEX
MEDULLA

PANCREAS
(*Islets of Langerhans*)

2 TESTES in Male
(2 OVARIES in Female
[plus placenta in pregnancy])

The glands of internal secretion are concerned with the CONTROL and CO-ORDINATION of processes which are widespread in the body — such as

<u>METABOLISM</u>

<u>GROWTH</u>

<u>Maintenance of Stability of INTERNAL ENVIRONMENT</u>

<u>RESISTANCE to STRESS</u>

and with

<u>REPRODUCTION</u>

Some other SYSTEMS also produce HORMONES which are passed into the BLOOD STREAM.
e.g.
STOMACH — GASTRIN

SMALL INTESTINE { ENTEROGASTRONE
SECRETIN
CHOLECYSTOKININ/
PANCREOZYMIN

KIDNEY { e.g. ERYTHROPOIETIN
RENIN
1:25 DHCC

All have a profuse BLOOD SUPPLY to receive their SECRETIONS

Although each ENDOCRINE GLAND has specific functions all are interdependent. Overactivity or Underactivity of one tends to affect the whole system.

87

THYROID

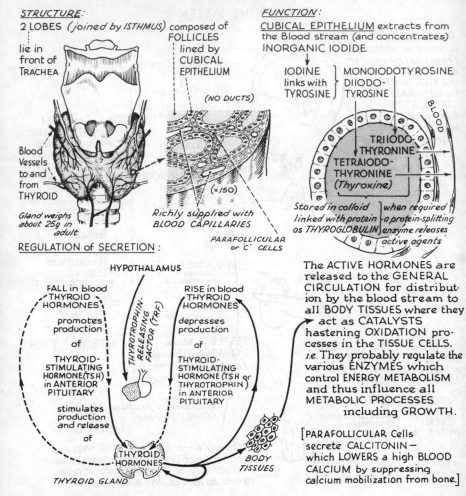

STRUCTURE:

2 LOBES *(joined by ISTHMUS)* composed of
FOLLICLES lined by CUBICAL EPITHELIUM

lie in front of TRACHEA

(NO DUCTS)

Blood Vessels to and from THYROID

Gland weighs about 25g in adult

Richly supplied with BLOOD CAPILLARIES

(×150)

PARAFOLLICULAR or 'C' CELLS

FUNCTION:

CUBICAL EPITHELIUM extracts from the Blood stream (and concentrates) INORGANIC IODIDE

IODINE links with TYROSINE } MONOIODOTYROSINE DIIODO-TYROSINE

TRIIODO-THYRONINE
TETRAIODO-THYRONINE
(Thyroxine)

BLOOD

Stored in colloid linked with protein as THYROGLOBULIN | *when required a protein-splitting enzyme releases active agents*

The ACTIVE HORMONES are released to the GENERAL CIRCULATION for distribution by the blood stream to all BODY TISSUES where they act as CATALYSTS hastening OXIDATION processes in the TISSUE CELLS. *i.e.* They probably regulate the various ENZYMES which control ENERGY METABOLISM and thus influence all METABOLIC PROCESSES including GROWTH.

[PARAFOLLICULAR Cells secrete CALCITONIN — which LOWERS a high BLOOD CALCIUM by suppressing calcium mobilization from bone.]

REGULATION of SECRETION:

HYPOTHALAMUS

FALL in blood THYROID HORMONES
promotes production of THYROID-STIMULATING HORMONE(TSH) in ANTERIOR PITUITARY
stimulates production and release of

THYROTROPHIN-RELEASING FACTOR (TRF)

RISE in blood THYROID HORMONES
depresses production of THYROID-STIMULATING HORMONE (TSH or THYROTROPHIN) in ANTERIOR PITUITARY

THYROID HORMONES

BODY TISSUES

THYROID GLAND

THYROID

UNDERACTIVITY

If the Thyroid shows *atrophy* of its secretory cells or is inadequately stimulated by the Anterior Pituitary, insufficient Hormonal Secretion is released to the blood stream and the rate at which cells use energy is reduced → Basal Metabolic Rate falls → less heat is produced → Body Temperature falls (and person feels cold). Energy stores increase (*e.g.* GLYCOGEN and FAT).

MYXOEDEMA results in the Adult

SLOWING UP OF ALL BODILY PROCESSES

Appetite is reduced; Weight increases. Gut movements are sluggish → Constipation. Heart and Respiratory Rates and Blood Pressure reduced. Thought processes slow down → Lethargy; Apathy. *SKIN*–Thick, leathery, puffy. *HAIR*–Brittle, sparse, dry. Husky voice. Blood cholesterol increases.

In the Child, congenital absence of the gland → *CRETINISM*

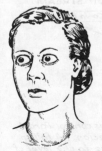

GROSS DWARFING. FAILURE of SKELETAL, SEXUAL, MENTAL GROWTH and DEVELOPMENT. All "milestones" of babyhood are delayed. N.B. *Coarse skin and hair; protruding tongue.*

Thyroxine (taken by mouth) restores individual to normal.

OVERACTIVITY

If an *enlarged* Thyroid shows increased activity of its secretory cells, excess Thyroid Hormones are distributed by the blood stream to the tissues of the body → speed up oxidations in the cells, i.e. rate at which all cells use energy → the Basal Metabolic Rate is raised. As a by-product of this increased cellular activity more heat is produced → rise in Body Temperature (person feels warm). → profuse sweating. Energy stores of body (i.e GLYCOGEN and FAT) are depleted.

SPEEDING UP OF ALL BODILY PROCESSES

{
Appetite increases but weight falls.
Movements of digestive tract are increased → Diarrhoea.
Heart and Respiratory Rates rise.
Blood Pressure is raised.
Muscular tremor and nervousness are marked.
Person becomes excitable and apprehensive.
}

[EXOPHTHALMOS (protrusion of eyeballs) may be due to an excess of some Pituitary Hormone. It is not due to an excess of Thyroid Hormones.]

Surgical removal of part of the overactive gland reduces Thyroid activity.

PARATHYROIDS

FOUR small glands composed of *cords of cells* which secrete Parathyroid
Hormone – **PARATHORMONE**

↓ or PTH

CAPILLARIES

↓

GENERAL CIRCULATION

↓

to ALL TISSUES of the body

But not all tissues are sensitive to it.

It plays an important rôle in CALCIUM and PHOSPHATE METABOLISM

PARATHYROID GLANDS

Situated behind THYROID

OESOPHAGUS

TRACHEA

×200

Function of EOSINOPHIL cells is unknown

Each weighs from 20-50mg in Adult

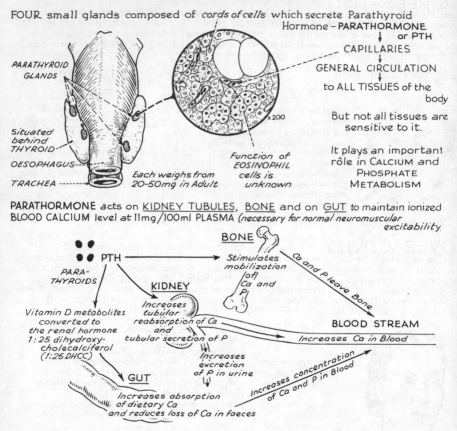

PARATHORMONE acts on <u>KIDNEY TUBULES</u>, <u>BONE</u> and on <u>GUT</u> to maintain ionized BLOOD CALCIUM level at 11mg/100ml PLASMA *(necessary for normal neuromuscular excitability*

BONE

PTH → Stimulates mobilization of Ca and P

PARA-THYROIDS

Ca and P leave Bone

KIDNEY

Increases tubular reabsorption of Ca and tubular secretion of P

Vitamin D metabolites converted to the renal hormone 1:25 dihydroxy-cholecalciferol (1:25 DHCC)

BLOOD STREAM

Increases Ca in Blood

Increases excretion of P in urine

GUT

Increases absorption of dietary Ca and reduces loss of Ca in faeces

Increases concentration of Ca and P in Blood

Alterations in concentration of CALCIUM ions in extracellular fluids control Parathyroid activity. A rise in blood Calcium depresses Parathyroid secretion. A fall in Calcium increases Parathyroid secretion. [Note: This response to the level of plasma Ca differs from calcitonin's – see 'thyroid'.]

PARATHYROIDS

UNDERACTIVITY

– Atrophy or removal of Parathyroid tissue causes fall in BLOOD CALCIUM level and increased excitability of Neuro-muscular tissue. This leads to severe convulsive disorder – *TETANY*.
Usual manifestations:- TWITCHINGS, NERVOUSNESS, OCCASIONAL SPASMS OF FACIAL AND LIMB MUSCLES.

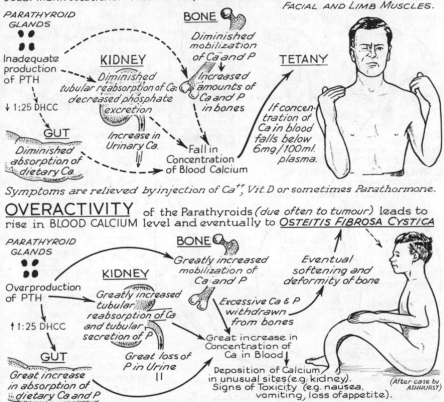

PARATHYROID GLANDS

Inadequate production of PTH

↓ 1:25 DHCC

KIDNEY
Diminished tubular reabsorption of Ca: decreased phosphate excretion

Increase in Urinary Ca.

GUT
Diminished absorption of dietary Ca.

BONE
Diminished mobilization of Ca and P

Increased amounts of Ca and P in bones

Fall in Concentration of Blood Calcium

TETANY

If concentration of Ca in blood falls below 6mg/100ml. plasma.

Symptoms are relieved by injection of Ca⁺⁺, Vit.D or sometimes Parathormone.

OVERACTIVITY

of the Parathyroids *(due often to tumour)* leads to rise in BLOOD CALCIUM level and eventually to <u>OSTEITIS FIBROSA CYSTICA</u>

PARATHYROID GLANDS

Overproduction of PTH

↑ 1:25 DHCC

GUT
Great increase in absorption of dietary Ca and P

KIDNEY
Greatly increased tubular reabsorption of Ca and tubular secretion of P

Great loss of P in Urine

BONE
Greatly increased mobilization of Ca and P

Excessive Ca & P withdrawn from bones

Great increase in Concentration of Ca in Blood

Eventual softening and deformity of bone

Deposition of Calcium in unusual sites (e.g. kidney). Signs of Toxicity (e.g. nausea, vomiting, loss of appetite).

(After case by ASHHURST)

The increased level of blood calcium eventually leads to excessive loss of CALCIUM in URINE and also of WATER since the salts are excreted in solution.
Excision of the overactive Parathyroid tissue abolishes syndrome.

91

ADRENAL GLANDS

There are TWO Adrenal Glands. They lie close to the kidneys.

Each has an outer CORTEX and an inner MEDULLA

RIGHT KIDNEY

LEFT KIDNEY

CAPSULE
ZONA GLOMERULOSA
ZONA FASCICULATA
ZONA RETICULARIS
MEDULLA

CORTEX

×80

Secretion is under control of the ANTERIOR PITUITARY ADRENO-CORTICOTROPHIC HORMONE (CORTICOTROPHIN, ACTH)

STRESS acts via HYPOTHALAMUS

FALL in blood CORTICOIDS

promotes production

of

CORTICOTROPHIN

stimulates production and release

of

CORTICOIDS

CORTICOTROPHIN release factor

RISE in blood CORTICOIDS

depresses production

of

CORTICOTROPHIN

This reciprocal relationship between A.P. and S.R. Cortex leads to balanced effects on →

ADRENAL GLAND

Production of the most powerful Salt-retaining Hormone – ALDOSTERONE – in the Zona Glomerulosa follows a reduction in blood sodium or increase in potassium or a drop in the volume of blood or body fluids (e.g. after haemorrhage, excess vomiting or diarrhoea).

The Adrenal Cortex secretes steroid HORMONES into the BLOOD STREAM. These travel to all tissues of the body.
The Adrenal Steroids share the same functions but to varying degrees:—

1. "MINERALOCORTICOID" EFFECT: (especially Aldosterone) — chief action on KIDNEY TUBULES

KIDNEY

Promotes RETENTION of SODIUM (with water)
Stimulates EXCRETION of POTASSIUM

2. "GLUCOCORTICOID" EFFECT: (Cortisol, Corticosterone) – on LIVER

LIVER

FORMATION of SUGAR from PROTEIN
(*This gives increased BLOOD SUGAR; promotes LIPOLYSIS*)
Corticoids also have ANTI-INFLAMMATORY and ANTI-ALLERGIC properties and stimulate r.b.c. formation.

SALT and WATER metabolism [Electrolyte balance]

CARBOHYDRATE metabolism

SEX HORMONES

ANDROGENS and OESTROGEN are found in extracts of the gland. They may represent stages in the formation of the other corticoids.

*The Adrenal Cortex is **essential to life** and plays an important rôle in states of stress.*

ADRENAL CORTEX

UNDERACTIVITY
Atrophy of the Adrenal Cortex (*occasionally occurs with destructive disease, e.g. T.B.*) gives **ADDISON'S DISEASE**

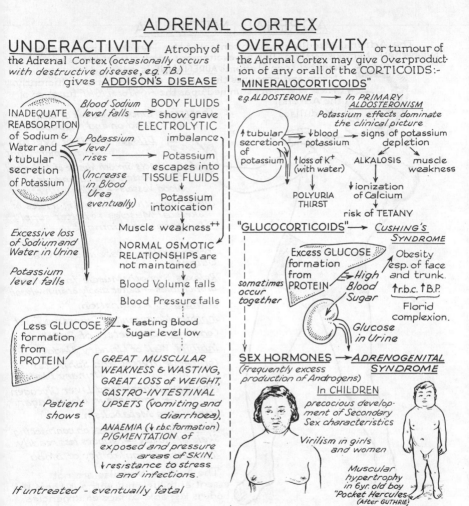

INADEQUATE REABSORPTION of Sodium & Water and ↓ tubular secretion of Potassium

Blood Sodium level falls → BODY FLUIDS show grave ELECTROLYTIC imbalance

Potassium level rises

(*Increase in Blood Urea eventually*)

→ Potassium escapes into TISSUE FLUIDS
↓
Potassium intoxication

Muscle weakness++

NORMAL OSMOTIC RELATIONSHIPS are not maintained

Blood Volume falls

Blood Pressure falls

Excessive loss of Sodium and Water in Urine

Potassium level falls

Less GLUCOSE formation from PROTEIN

- - Fasting Blood Sugar level low

Patient shows

GREAT MUSCULAR WEAKNESS & WASTING, GREAT LOSS of WEIGHT, GASTRO-INTESTINAL UPSETS (vomiting and diarrhoea), ANAEMIA (↓ r.b.c. formation) PIGMENTATION of exposed and pressure areas of SKIN. ↓ resistance to stress and infections.

If untreated - eventually fatal

Administration of *Adrenal Hormones* and salt restores individual to normal.

OVERACTIVITY
or tumour of the Adrenal Cortex may give Overproduction of any or all of the CORTICOIDS:-

"MINERALOCORTICOIDS"

e.g. ALDOSTERONE → *In PRIMARY ALDOSTERONISM*
Potassium effects dominate the clinical picture

↑tubular secretion of potassium → ↓blood potassium → signs of potassium depletion

↑loss of K^+ (with water) → ALKALOSIS   muscle weakness

POLYURIA THIRST

↓ionization of Calcium
↓
risk of TETANY

"GLUCOCORTICOIDS" → *CUSHING'S SYNDROME*

Excess GLUCOSE formation from PROTEIN

High Blood Sugar

sometimes occur together

Obesity esp. of face and trunk.
↑r.b.c. ↑B.P.
Florid complexion.

Glucose in Urine

SEX HORMONES → *ADRENOGENITAL SYNDROME*
(Frequently excess production of Androgens)

In CHILDREN precocious development of Secondary Sex characteristics

Virilism in girls and women

Muscular hypertrophy in 6 yr. old boy "Pocket Hercules" (After GUTHRIE)

Removal of the over-secreting tissue or tumour restores individual.

ADRENAL MEDULLA

The Adrenal Medulla is under control of the Hypothalamus via the Sympathetic Nervous System. During excitement or circumstances which demand special efforts __ADRENALINE__ is released into the blood stream and prepares the various systems to react efficiently. (These effects are summed up as the *'FIGHT or FLIGHT'* function of the Adrenal Medullae.)

It <u>Constricts</u> Smooth Muscle of Skin → Hairs 'stand on end'; 'Gooseflesh'.

<u>Dilates</u> Pupil of Eye to admit more light.

<u>Constricts</u> Smooth Muscle of Abdominal Blood Vessels and Cutaneous Blood Vessels → Pallor with Fright.

<u>Dilates</u> Smooth Muscle in Arterioles of Skeletal Muscles i.e better supply to organs requiring it in emergency.

<u>Increases</u> Heart Rate and Cardiac Output.

<u>Relaxes</u> Smooth Muscle in Wall of Bronchioles → better supply of air to alveoli.

<u>Stimulates</u> Respiration.

<u>Inhibits</u> Movements of Digestive Tract.

<u>Contracts</u> Sphincters of Gut.

<u>Inhibits</u> Wall of Urinary Bladder.

<u>Contracts</u> Ureters and Sphincter of Urinary Bladder.

<u>Mobilizes</u> Muscle and Liver Glycogen → increase in Blood Sugar.

<u>Stimulates</u> Metabolism.

<u>Exerts favourable effect</u> on contracting Skeletal Muscle → Fatigues less readily.

<u>Increases</u> Coagulability of Blood.

The Adrenal Medullae are not essential to life – but without them the body is less able to face emergencies.

Stimulation of another part of Hypothalamus leads to release of NORADRENALINE (from Adrenal Medullae) → Vasoconstriction → Rise in B.P.

ANTERIOR PITUITARY

This is the MASTER GLAND of the ENDOCRINE SYSTEM.
It regulates the activity of the other Endocrine
Glands, including the GONADS, and influences
ALL METABOLIC PROCESSES
including GROWTH.

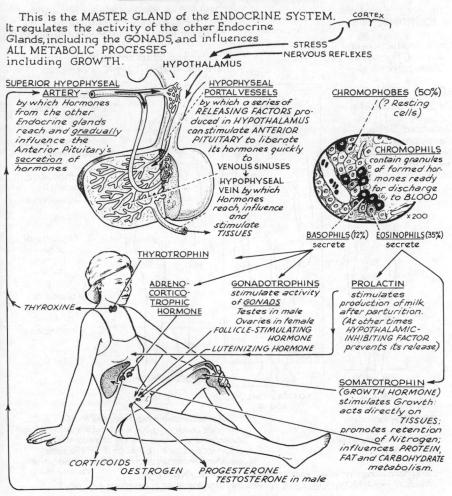

CORTEX
STRESS
NERVOUS REFLEXES
HYPOTHALAMUS

SUPERIOR HYPOPHYSEAL
ARTERY —
by which Hormones
from the other
Endocrine glands
reach and gradually
influence the
Anterior Pituitary's
secretion of
hormones

HYPOPHYSEAL
PORTAL VESSELS
by which a series of
RELEASING FACTORS pro-
duced in HYPOTHALAMUS
can stimulate ANTERIOR
PITUITARY to liberate
its hormones quickly
to
VENOUS SINUSES

HYPOPHYSEAL
VEIN by which
Hormones
reach, influence
and
stimulate
TISSUES

CHROMOPHOBES (50%)
(? Resting
cells)

CHROMOPHILS
contain granules
of formed hor-
mones ready
for discharge
to BLOOD
x 200

BASOPHILS (12%)
secrete

EOSINOPHILS (35%)
secrete

THYROTROPHIN

THYROXINE

ADRENO-
CORTICO-
TROPHIC
HORMONE

GONADOTROPHINS
stimulate activity
of GONADS
Testes in male
Ovaries in female
FOLLICLE-STIMULATING
HORMONE
LUTEINIZING HORMONE

PROLACTIN
stimulates
production of milk
after parturition.
(At other times
HYPOTHALAMIC-
INHIBITING FACTOR
prevents its release)

SOMATOTROPHIN
(GROWTH HORMONE)
stimulates Growth:
acts directly on
TISSUES;
promotes retention
of Nitrogen;
influences PROTEIN,
FAT and CARBOHYDRATE
metabolism.

CORTICOIDS
OESTROGEN
PROGESTERONE
TESTOSTERONE in male

95

UNDERACTIVITY of ANTERIOR PITUITARY

Deficiency or absence
of EOSINOPHIL cells

↓

Underproduction of
GROWTH Hormone
(Somatotrophin)

↓

LORAIN DWARF

Delayed Skeletal
Growth and
Retarded Sexual
Development
but alert, intelligent,
well proportioned
child.

Destructive disease of part of Anterior
Pituitary (usually with damage to
Posterior Pituitary and/or Hypothalamus)

↓

Underproduction of GROWTH and other
ENDOCRINE-TROPHIC Hormones

↓

FRÖHLICH'S DWARF

Stunting of Growth,
Obesity (*Large
appetite for sugar*);
Arrested Sexual
Development;
Lethargic;
Somnolent;
Mentally
Subnormal.

If Atrophy of
other
Endocrine
glands

↓

Signs of
deficiency
of their
hormones.

AGE 13

NORMAL CHILD
AGE 13

AGE 13

*GROWTH hormone
restores growth and
development pattern
to normal.*

*A similar condition occurs in ADULT
without dwarfing but with suppression of sex
functions and regression of secondary sex
characteristics. Growth and Gonado-
trophic hormones aid in restoring patient
to normal.*

*Complete atrophy of all secreting cells of Ant. Pituitary
in adult →* SIMMOND's DISEASE *(Features are those of premature senility.)*

OVERACTIVITY of PITUITARY CELLS

EOSINOPHILS

Overactivity (or tumour) of **EOSINOPHILS** leads to Overproduction of GROWTH Hormone → **GIANTISM** in Child ; **ACROMEGALY** in Adult.

Overgrowth of all Body Tissues

Onset before bony epiphyses have closed at puberty

Onset after puberty

Increased Nitrogen retention. (Protein, Carbohydrate and Fat Metabolism of ALL CELLS of the body affected)

Bones thicken esp. of FACE, JAW, NOSE, HANDS and FEET

Coarse thick SKIN

Overgrowth of SOFT TISSUES and INTERNAL ORGANS (e.g. Heart, Spleen, Stomach, etc.)

Long bones grow in length (Height 7-8 feet). Overgrowth of MUSCLES

Due to the diabetogenic action of the hormone these patients have a HIGH BLOOD SUGAR with SUGAR in the URINE. Fat is mobilized and free fatty acids are used for energy instead of glucose. Other features of this condition are often due to pressure of tumour on surrounding brain tissue or sometimes to overproduction of other Anterior Pituitary Hormones.

AGE 13

Destruction of the overactive tissue—usually by RADIUM therapy — prevents progression of the condition.

BASOPHILS

Overproduction esp. of ACTH ↓ (Corticotrophin)
Overstimulation of Adrenal Cortex

CUSHING'S SYNDROME

↓ absorption of Ca++ from gut → osteoporosis

Tissue proteins ↓ depleted
muscle wasting weakness:
atrophied skin purple striae

High blood sugar obesity
sugar in urine polyuria thirst

↑ Glucocorticoid effect → ↑BP ↑r.b.c.
Florid complexion

Excess Corticoids

↑Mineralocorticoid effect is corrected by the body's adjustments to aldosterone.

Androgens
Hirsutism and suppression of menstrual cycle.

↑Anti-inflammatory and anti-allergic effects: altered resistance to stress.

infections spread readily, wounds heal slowly, normal allergic reactions are prevented.

This condition also results from overactivity of Adrenal Cortex itself.

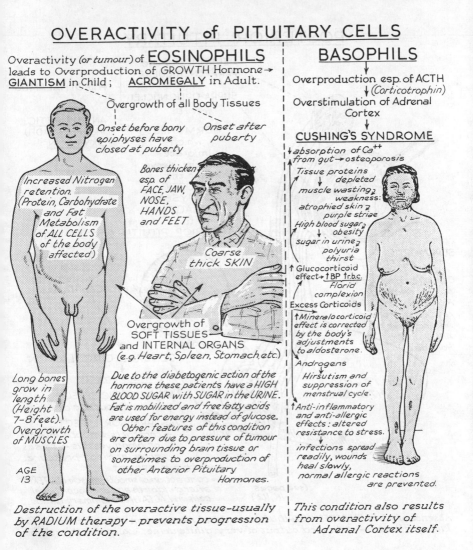

POSTERIOR PITUITARY

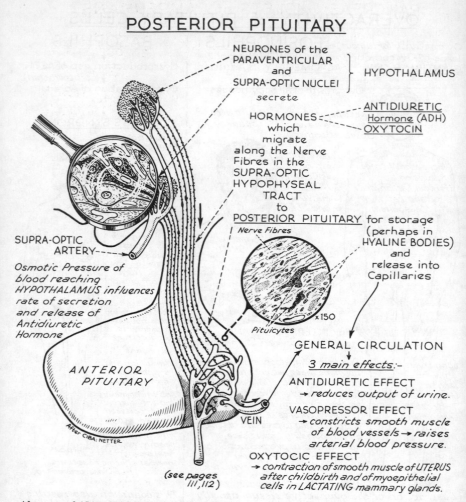

NEURONES of the
PARAVENTRICULAR
and
SUPRA-OPTIC NUCLEI } HYPOTHALAMUS
secrete

HORMONES
which
migrate
along the Nerve
Fibres in the
SUPRA-OPTIC
HYPOPHYSEAL
TRACT
to

ANTIDIURETIC
Hormone (ADH)
OXYTOCIN

POSTERIOR PITUITARY for storage
(perhaps in
HYALINE BODIES)
and
release into
Capillaries

Nerve Fibres

×150

Pituicytes

SUPRA-OPTIC
ARTERY

Osmotic Pressure of
blood reaching
HYPOTHALAMUS influences
rate of secretion
and release of
Antidiuretic
Hormone

ANTERIOR
PITUITARY

After CIBA: NETTER

VEIN

GENERAL CIRCULATION

3 main effects:-

ANTIDIURETIC EFFECT
→ reduces output of urine.

VASOPRESSOR EFFECT
→ constricts smooth muscle
of blood vessels → raises
arterial blood pressure.

OXYTOCIC EFFECT
→ contraction of smooth muscle of UTERUS
after childbirth and of myoepithelial
cells in LACTATING mammary glands.

(see pages
111, 112)

Absence of ADH → DIABETES INSIPIDUS (diminished reabsorption of water from
kidney filtrate → increased output of very dilute urine; excessive thirst).

PANCREAS: ISLETS of LANGERHANS

ISLETS of LANGERHANS make up 1-2% of Pancreatic Tissue.

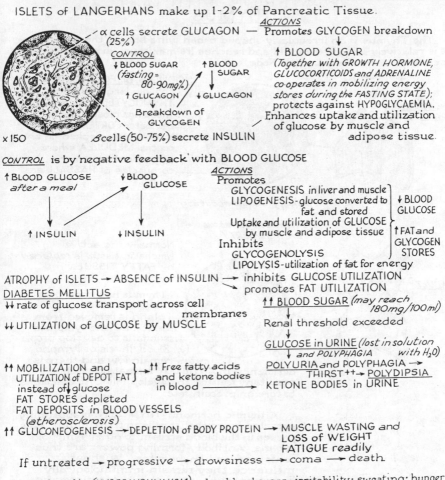

ACTIONS

α cells secrete GLUCAGON — Promotes GLYCOGEN breakdown
(25%)
    ↓
CONTROL   ↑ BLOOD SUGAR
↓ BLOOD SUGAR   ↑ BLOOD   *(Together with GROWTH HORMONE,*
(fasting =   SUGAR   *GLUCOCORTICOIDS and ADRENALINE*
80-90mg%)   ↓   *co-operates in mobilizing energy*
↓   *stores during the FASTING STATE);*
↑ GLUCAGON   ↓ GLUCAGON   protects against HYPOGLYCAEMIA.
↓   Enhances uptake and utilization
Breakdown of   of glucose by muscle and
GLYCOGEN   adipose tissue.

×150   β cells (50-75%) secrete INSULIN

CONTROL is by 'negative feedback' with BLOOD GLUCOSE

ACTIONS
↑ BLOOD GLUCOSE   ↓ BLOOD   Promotes
after a meal   GLUCOSE   GLYCOGENESIS in liver and muscle ⎤
    LIPOGENESIS - glucose converted to   ⎟ ↓ BLOOD
    fat and stored   ⎬ GLUCOSE
↓   ↓   Uptake and utilization of GLUCOSE   ⎟
↑ INSULIN   ↓ INSULIN   by muscle and adipose tissue ⎦ ↑ FAT and
    Inhibits   GLYCOGEN
    GLYCOGENOLYSIS   STORES
    LIPOLYSIS - utilization of fat for energy

ATROPHY of ISLETS → ABSENCE of INSULIN → inhibits GLUCOSE UTILIZATION
                  → promotes FAT UTILIZATION
DIABETES MELLITUS
↓↓ rate of glucose transport across cell   ↑↑ BLOOD SUGAR *(may reach*
             membranes   *180mg/100ml)*
↓↓ UTILIZATION of GLUCOSE by MUSCLE   Renal threshold exceeded
    ↓
    GLUCOSE in URINE *(lost in solution*
    ↓ *and POLYPHAGIA   with H₂O)*

    POLYURIA and POLYPHAGIA →
↑↑ MOBILIZATION and ⎤ ↑↑ Free fatty acids   THIRST ++→ POLYDIPSIA
UTILIZATION of DEPOT FAT ⎦ and ketone bodies   KETONE BODIES in URINE
instead of ↓ glucose   in blood
FAT STORES depleted
FAT DEPOSITS in BLOOD VESSELS
(atherosclerosis)
↑↑ GLUCONEOGENESIS → DEPLETION of BODY PROTEIN → MUSCLE WASTING and
                            LOSS of WEIGHT
                            FATIGUE readily
    If untreated → progressive → drowsiness ——→ coma → death.

Excess Insulin (HYPERINSULINISM) → low blood sugar → irritability; sweating; hunger.
If untreated → reduction of metabolism of nervous tissues → giddiness → coma → death.

THYMUS

The Thymus is an irregularly-shaped organ lying behind the breast bone. It is relatively large in the child and reaches its maximum size at puberty. It closely resembles a Lymph Node.

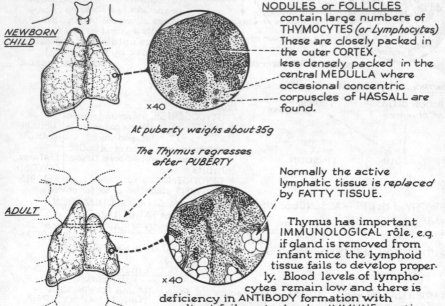

NEWBORN CHILD

×40

ADULT

×40

At puberty weighs about 35g

The Thymus regresses after PUBERTY

NODULES or FOLLICLES
contain large numbers of THYMOCYTES (or Lymphocytes) These are closely packed in the outer CORTEX, less densely packed in the central MEDULLA where occasional concentric corpuscles of HASSALL are found.

Normally the active lymphatic tissue is *replaced* by FATTY TISSUE.

Thymus has important IMMUNOLOGICAL rôle, e.g. if gland is removed from infant mice the lymphoid tissue fails to develop properly. Blood levels of lymphocytes remain low and there is deficiency in ANTIBODY formation with resultant failure to develop IMMUNE reactions to foreign protein, etc.

A thymic hormone-THYMOSIN-is apparently secreted by epithelial cells in the thymus. This passes in the blood stream to other lymphoid organs. Antibody-forming powers are thus conferred on lymphocytes. Without its influence they remain immunologically 'incompetent' — i.e. unable to form specific antibodies.

100

CHAPTER 7.
REPRODUCTIVE SYSTEM —— MALE

PRIMARY SEX ORGANS *produce the MALE GERM CELLS – <u>SPERMATOZOA</u>*
TESTES (Two) *and the MALE SEX HORMONE – <u>TESTOSTERONE</u>*

Testosterone is responsible for
development at Puberty of :-
<u>SECONDARY SEX ORGANS</u>
EPIDIDYMIS (Two) ——— *transfer Spermatozoa from the Testes.*
VAS DEFERENS (Two) ———
SEMINAL VESICLES (Two) -} *secrete fluid medium for transport of Spermatozoa.*
PROSTATE GLAND ———
PENIS ——————— *transfers Spermatozoa from male to female.*

and
appearance of
<u>SECONDARY SEX CHARACTERISTICS</u>
Laryngeal changes → Deep voice.
Pubic, Axillary and Facial hair.
Characteristic Male shape of
body.

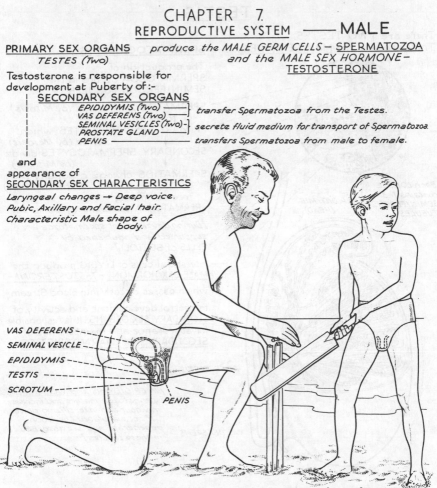

VAS DEFERENS ---
SEMINAL VESICLE ---
EPIDIDYMIS ---
TESTIS ---
SCROTUM ---
PENIS

In the male the process of spermatogenesis starts just after
puberty and is normally continuous until old age.

TESTIS

There are TWO TESTES.
These produce the <u>Male GERM CELLS</u>:–

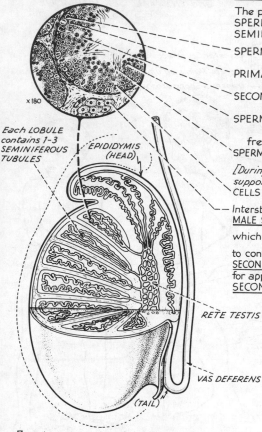

x 180

*Each LOBULE
contains 1-3
SEMINIFEROUS
TUBULES*

*EPIDIDYMIS
(HEAD)*

RETE TESTIS

VAS DEFERENS

(TAIL)

SPERMATOGENESIS

The production of
SPERMATOZOA occurs in the
SEMINIFEROUS TUBULES.

SPERMATOGONIA divide *(by Mitosis)*
    to form
PRIMARY SPERMATOCYTES which
      divide *(by Meiosis)*
SECONDARY SPERMATOCYTES divide
      (by Mitosis)
SPERMATIDS change gradually *(no
      further division)*
  free swimming
SPERMATOZOA *(approximately 0·1mm long)*

*[During development spermatozoa are
supported and nourished by*
CELLS of SERTOLI.*]*

Interstitial cells of LEYDIG produce the
<u>MALE SEX HORMONE</u> – <u>*TESTOSTERONE*</u> –

which passes directly into Blood Stream

to control development and activity of
<u>SECONDARY SEX ORGANS</u>. It is responsible
for appearance and maintenance of
<u>SECONDARY SEX CHARACTERISTICS</u>.

*[In <u>EPIDIDYMIS</u> and <u>VAS DEFERENS</u>
spermatozoa mature and increase
in vigour and in fertilizing power.
If not ejaculated they soon
degenerate and are absorbed
in these tubules.]*

*Events occurring in the testes are under CONTROL of HORMONES chiefly
those of the HYPOTHALAMUS and ANTERIOR PITUITARY.*

MALE SECONDARY SEX ORGANS

These are the organs adapted for TRANSFER of live SPERMATOZOA from male to female.

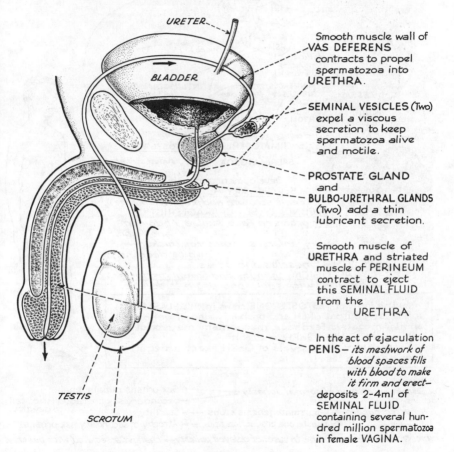

URETER

BLADDER

Smooth muscle wall of VAS DEFERENS contracts to propel spermatozoa into URETHRA.

SEMINAL VESICLES (Two) expel a viscous secretion to keep spermatozoa alive and motile.

PROSTATE GLAND and BULBO-URETHRAL GLANDS (Two) add a thin lubricant secretion.

Smooth muscle of URETHRA and striated muscle of PERINEUM contract to eject this SEMINAL FLUID from the URETHRA

In the act of ejaculation PENIS— *its meshwork of blood spaces fills with blood to make it firm and erect*— deposits 2-4ml of SEMINAL FLUID containing several hundred million spermatozoa in female VAGINA.

TESTIS

SCROTUM

CONTROL OF EVENTS IN THE TESTIS

Between the ages of 13 and 16 years the hypothalamus begins to secrete GONADOTROPHIN-RELEASING HORMONE (GnRH).

Seminiferous Tubules

Cells of Sertoli   Cells of Leydig

GnRH (−) (−) (−) (−)

LH (Luteinising hormone)

FSH (Follicle-stimulating hormone)

TESTIS

SEMINIFEROUS TUBULES ↓ (Androgen-binding protein)

INTERSTITIAL CELLS of LEYDIG

INHIBIN to blood stream

Cells of SERTOLI support germ cells → ABP

SPERMATOGENESIS   TESTOSTERONE

SPERMATOZOA   Blood stream

Development and maintenance of
SECONDARY SEX ORGANS
(Penis, scrotum and prostate enlarge)
SECONDARY SEX CHARACTERISTICS
(Hair grows on Face, Axillae, Abdomen and Pubis:
Larynx enlarges, vocal cords thicken →
Voice deepens.)

GROWTH of muscles and bones.
PSYCHOLOGICAL changes and Sexual Instincts
associated with adulthood begin to develop.

As the level of TESTOSTERONE rises in the PLASMA, it inhibits (−) output of LH and probably also of GnRH. A similar negative feedback mechanism may exist between INHIBIN, GnRH and FSH.

Note that rising levels of GnRH exert a negative feedback on itself.

If Atrophy of Testes occurs   at time of normal puberty → Sex organs remain small.
                                → Secondary sex characteristics fail to develop.
        after puberty → Spermatogenesis stops → Sterility.
                → Testosterone production falls → Atrophy of secondary sex organs.

Injections of Testosterone in cases of delayed puberty → changes associated with puberty.

104

REPRODUCTIVE SYSTEM — FEMALE

PRIMARY SEX ORGANS
OVARIES (Two)

*produce the FEMALE GERM CELLS — <u>OVA</u>
and the FEMALE SEX HORMONES —
<u>OESTROGEN</u> and <u>PROGESTERONE</u>*

Oestrogen and Progesterone are responsible
for development at Puberty of:-

<u>SECONDARY SEX ORGANS</u>

FALLOPIAN TUBES (Two) — for the transfer of the Ova from Ovaries.
VAGINA ———————— for the reception of the Male Germ Cells.
*UTERUS ———————— for the nutrition and development of the
                  fertilized Egg Cell → developing embryo.*
MAMMARY GLANDS (Two) - for the nutrition of the New Individual after birth.

and
appearance and
maintenance of
<u>SECONDARY SEX
CHARACTERISTICS</u>

*Development of
   Breasts*

*Axillary and Pubic
        hair*

*Typical Feminine
Proportions of body*

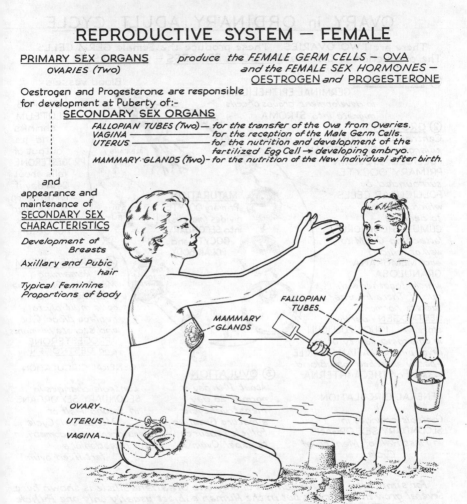

FALLOPIAN
TUBES

MAMMARY
GLANDS

OVARY
UTERUS
VAGINA

*In the female the cyclical production of ova starts just after puberty and
continues (unless interrupted by pregnancy or disease) until the menopause.*

OVARY in ORDINARY ADULT CYCLE

There are TWO OVARIES. These produce the Female GERM CELLS.
The production of OVA is a **CYCLICAL PROCESS** — **OOGENESIS**

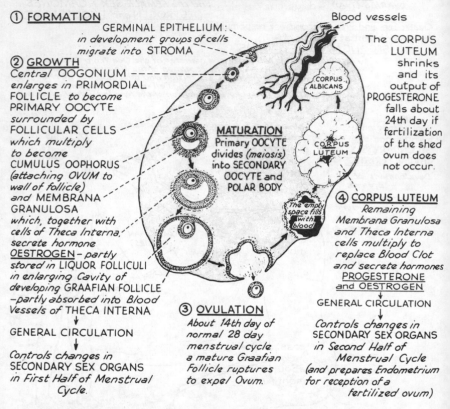

① **FORMATION**

GERMINAL EPITHELIUM:
*in development groups of cells
migrate into STROMA*

Blood vessels

② **GROWTH**
Central OOGONIUM
enlarges in PRIMORDIAL
FOLLICLE *to become*
PRIMARY OOCYTE
surrounded by
FOLLICULAR CELLS
*which multiply
to become*
CUMULUS OOPHORUS
*(attaching OVUM to
wall of follicle)
and* MEMBRANA
GRANULOSA
*which, together with
cells of Theca Interna,
secrete hormone*
OESTROGEN – *partly
stored in* LIQUOR FOLLICULI
*in enlarging Cavity of
developing* GRAAFIAN FOLLICLE
*–partly absorbed into Blood
Vessels of* THECA INTERNA

↓

GENERAL CIRCULATION

↓

Controls changes in
SECONDARY SEX ORGANS
*in First Half of Menstrual
Cycle.*

CORPUS ALBICANS

CORPUS LUTEUM

MATURATION
Primary OOCYTE
divides (meiosis)
into SECONDARY
OOCYTE and
POLAR BODY

The empty
space fills
with
blood

③ **OVULATION**
*About 14th day of
normal 28 day
menstrual cycle
a mature Graafian
Follicle ruptures
to expel Ovum.*

The CORPUS
LUTEUM
shrinks
and its
output of
PROGESTERONE
falls about
24th day if
fertilization
of the shed
ovum does
not occur.

④ **CORPUS LUTEUM**
*Remaining
Membrana Granulosa
and Theca Interna
cells multiply to
replace Blood Clot
and secrete hormones*
PROGESTERONE
and **OESTROGEN**

↓

GENERAL CIRCULATION

↓

Controls changes in
SECONDARY SEX ORGANS
*in Second Half of
Menstrual Cycle
(and prepares Endometrium
for reception of a
fertilized ovum)*

*For simplicity the development of only one Graafian Follicle is shown here.
Several grow in each cycle but in the Human subject usually only one Follicle
ruptures. The others atrophy: i.e. ONE MATURE OVUM is shed each month.
Events in the OVARY are under control of HYPOTHALAMIC RELEASING FACTOR
and ANTERIOR PITUITARY HORMONES.*

OVARY in PREGNANCY

When pregnancy occurs the ordinary ovarian cycle is suspended.

After the first 14 days the developing placenta secretes a **LUTEINIZING HORMONE** (chorionic gonadotrophin).

Under its influence the CORPUS LUTEUM continues to grow until it may come to occupy 30% to 50% of the total volume of the OVARY.

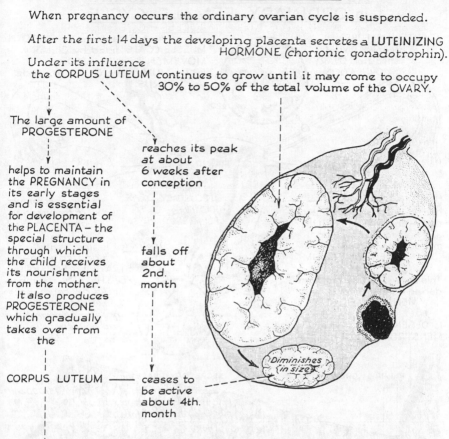

The large amount of PROGESTERONE

helps to maintain the PREGNANCY in its early stages and is essential for development of the PLACENTA – the special structure through which the child receives its nourishment from the mother.

It also produces PROGESTERONE which gradually takes over from the

CORPUS LUTEUM ——

reaches its peak at about 6 weeks after conception

falls off about 2nd. month

ceases to be active about 4th. month

Diminishes in size

PLACENTAL PROGESTERONE —— takes over to maintain the PREGNANCY, and to help to prepare the mammary glands for Lactation.

ADULT PELVIC SEX ORGANS
in ORDINARY FEMALE CYCLE

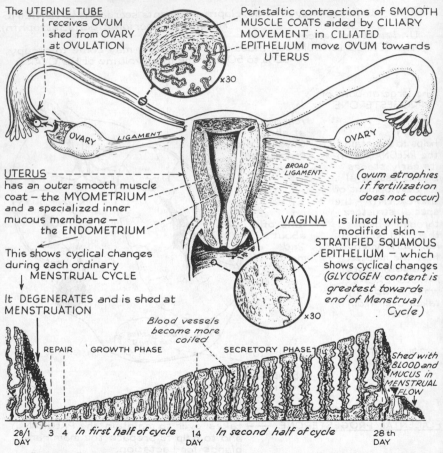

The UTERINE TUBE
  receives OVUM
  shed from OVARY
  at OVULATION

Peristaltic contractions of SMOOTH MUSCLE COATS aided by CILIARY MOVEMENT in CILIATED EPITHELIUM move OVUM towards UTERUS

×30

OVARY

LIGAMENT

OVARY

BROAD LIGAMENT

(ovum atrophies if fertilization does not occur)

UTERUS
has an outer smooth muscle coat – the MYOMETRIUM and a specialized inner mucous membrane – the ENDOMETRIUM

This shows cyclical changes during each ordinary MENSTRUAL CYCLE

It DEGENERATES and is shed at MENSTRUATION

VAGINA is lined with modified skin – STRATIFIED SQUAMOUS EPITHELIUM – which shows cyclical changes (GLYCOGEN content is greatest towards end of Menstrual Cycle)

×30

Blood vessels become more coiled

REPAIR   GROWTH PHASE   SECRETORY PHASE

Shed with BLOOD and MUCUS in MENSTRUAL FLOW

28/1 DAY   3 4   In first half of cycle   14 DAY   In second half of cycle   28th DAY

Rhythmical changes occur in UTERUS, UTERINE TUBES and VAGINA under action of OVARIAN Hormones. (see page 113)

UTERINE TUBES AND UTERUS
IN CYCLE ENDING IN PREGNANCY

The fimbriated end of the UTERINE TUBE receives OVUM at OVULATION.
The uterine tube also transmits SPERMATOZOA towards the OVA.

FERTILIZATION
–or fusion of OVUM and SPERM– occurs in outer third of uterine tube

CLEAVAGE
After fertilization in the uterine tube the fertilized Ovum or ZYGOTE undergoes several divisions

Blastocyst

IMPLANTATION
For a few days embryo gets OXYGEN and nutrients (*by diffusion*) from the uterine glandular secretion.

Embryo sticks to lining of WOMB. Its surface TROPHOBLAST cells fuse with, destroy and finally penetrate the ENDOMETRIUM (*now called the DECIDUA*).

Embryo now absorbs TISSUE FLUIDS and CELLULAR DEBRIS.

Spermatozoa are of two kinds — if ovum is fertilized by one kind the resulting embryo will be MALE; if by the other, FEMALE.
(All other sperms present atrophy)

Ciliary currents and PERISTALTIC contractions in Uterine Tube carry BLASTOCYST into Uterine Secretion about 4th–7th day

Outgrowths from outer layer of Blastocyst begin to invade Endometrium

The Endometrium is in LUTEAL PHASE and continues to grow. (*No menstrual degeneration occurs*) Glands are actively secreting mucus.

CHORIONIC VILLI – finger-like projections from the embryo invade mother's endometrial blood vessels.

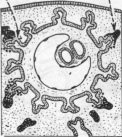

UTERUS

PLACENTATION

PROGESTERONE is necessary for the development of the PLACENTA - the special organ through which the developing child receives nourishment from the mother.

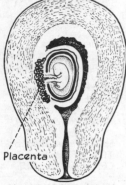

Placenta

Blood vessels develop in the Chorionic Villi which are now interlocked with mother's tissues and surrounded by mother's blood. Structure formed in this way is the PLACENTA

As PREGNANCY ADVANCES

PLACENTAL OESTROGEN, PROGESTERONE and GONADOTROPHINS in blood stream condition the maintenance of pregnancy

Umbilical Cord

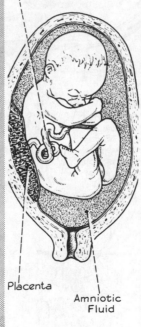

Placenta

Amniotic Fluid

PARTURITION

About 40 weeks after conception the process of <u>CHILDBIRTH</u> usually begins

<u>1ST STAGE</u> usually lasts up to 14 hours with a first birth

MYOMETRIUM
  Uterine muscle is now very greatly distended
↓
Rhythmic contractions increase in strength and frequency
↓
Press on Amniotic Fluid

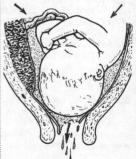

Neck of Womb (Cervix) and its opening (Os) dilate

<u>Membranes rupture</u> and <u>AMNIOTIC FLUID</u> escapes.

UTERUS

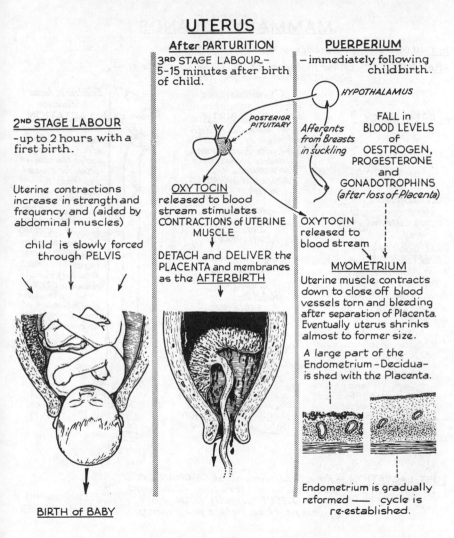

After PARTURITION

3RD STAGE LABOUR –
5-15 minutes after birth
of child.

PUERPERIUM

– immediately following
        childbirth.

HYPOTHALAMUS

POSTERIOR
PITUITARY

*Afferents
from Breasts
in suckling*

FALL in
BLOOD LEVELS
of
OESTROGEN,
PROGESTERONE
and
GONADOTROPHINS
(after loss of Placenta)

OXYTOCIN
released to blood
stream stimulates
CONTRACTIONS of UTERINE
MUSCLE

DETACH and DELIVER the
PLACENTA and membranes
as the <u>AFTERBIRTH</u>

OXYTOCIN
released to
blood stream

MYOMETRIUM

Uterine muscle contracts
down to close off blood
vessels torn and bleeding
after separation of Placenta.
Eventually uterus shrinks
almost to former size.

A large part of the
Endometrium – Decidua –
is shed with the Placenta.

Endometrium is gradually
reformed —— cycle is
re-established.

2ND STAGE LABOUR

–up to 2 hours with a
first birth.

Uterine contractions
increase in strength and
frequency and (aided by
abdominal muscles)

child is slowly forced
through PELVIS

<u>BIRTH of BABY</u>

MAMMARY GLANDS

<u>In CHILDHOOD</u> – *rudimentary ducts in fibrous tissue.*

<u>After PUBERTY</u> – *under influence of OESTROGEN, ducts grow;*
PROGESTERONE → acini sprout.

○ HYPOTHALAMUS

<u>In PREGNANCY</u>

<u>After CHILDBIRTH</u>
fall in OESTROGEN and PROGESTERONE (after loss of Placenta) inhibits secretion of HYPOTHALAMIC-PROLACTIN-INHIBITING FACTOR.

PROLACTIN is released by ANTERIOR PITUITARY

<u>PROGESTERONE</u>
from Corpus Luteum and
<u>OESTROGEN and PROGESTERONE</u>
from Placenta stimulate further GROWTH of DUCTS and ACINI.

stimulates prepared <u>GLAND ACINI</u> to secrete MILK

AREOLA and NIPPLE become darkly PIGMENTED

Constituents of MILK are derived from blood flowing through gland

HYPO-THALAMUS

Emotional factors influence <u>LACTATION</u>
Milk production starts 3-4 days after childbirth and is maintained (often for several months) by <u>PROLACTIN</u>.

PROLACTIN

OXYTOCIN

<u>AFFERENT</u> *nerve impulses, set up by Child* <u>SUCKLING</u> *give rise to reflex* <u>EFFERENT</u> *nerve impulses to Post. Pituitary for the release of* <u>OXYTOCIN</u> *– which stimulates the 'let-down' of milk from ducts to reservoirs.*

<u>POST-LACTATION</u> – *Re-establishment of OVARIAN CYCLE: regression of secreting ACINI and DUCT TISSUE till only little more present than before pregnancy.*

112

PITUITARY, OVARIAN and ENDOMETRIAL CYCLES

HYPOTHALAMUS secretes Gonadotrophin Releasing Factor into hypothalamic-hypophyseal portal circulation

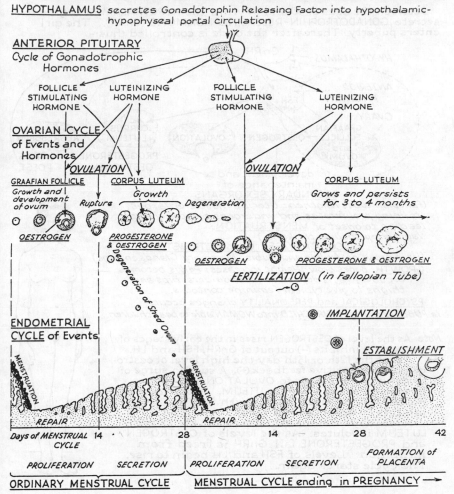

ANTERIOR PITUITARY
Cycle of Gonadotrophic Hormones

FOLLICLE STIMULATING HORMONE   LUTEINIZING HORMONE   FOLLICLE STIMULATING HORMONE   LUTEINIZING HORMONE

OVARIAN CYCLE
of Events and Hormones

OVULATION   _OVULATION_

GRAAFIAN FOLLICLE
Growth and development of ovum

CORPUS LUTEUM
Growth

Rupture   Degeneration

CORPUS LUTEUM
Grows and persists for 3 to 4 months

OESTROGEN   _PROGESTERONE & OESTROGEN_

Degeneration of Shed Ovum

OESTROGEN   _PROGESTERONE & OESTROGEN_

FERTILIZATION (in Fallopian Tube)

IMPLANTATION

ESTABLISHMENT

ENDOMETRIAL CYCLE of Events

MENSTRUATION   REPAIR   MENSTRUATION   REPAIR

Days of MENSTRUAL CYCLE   14   28   14   28   42

PROLIFERATION   SECRETION   PROLIFERATION   SECRETION   FORMATION of PLACENTA

ORDINARY MENSTRUAL CYCLE   MENSTRUAL CYCLE ending in PREGNANCY ⟶

113

CONTROL OF EVENTS IN THE OVARY

Between the ages of 10 and 14 years the HYPOTHALAMUS begins to secrete GONADOTROPHIN-RELEASING HORMONE (Gn.RH). The girl enters puberty. Thereafter the cycle is controlled thus:-

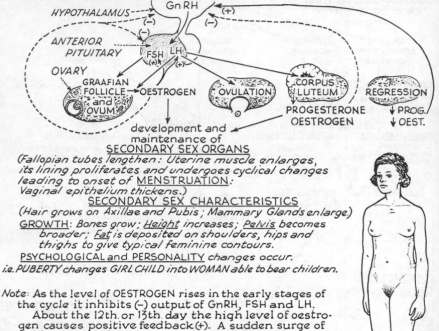

SECONDARY SEX ORGANS
(Fallopian tubes lengthen: Uterine muscle enlarges, its lining proliferates and undergoes cyclical changes leading to onset of MENSTRUATION:
Vaginal epithelium thickens.)

SECONDARY SEX CHARACTERISTICS
(Hair grows on Axillae and Pubis; Mammary Glands enlarge.)
GROWTH: Bones grow; Height increases; Pelvis becomes broader; Fat is deposited on shoulders, hips and thighs to give typical feminine contours.

PSYCHOLOGICAL and PERSONALITY changes occur.
i.e. PUBERTY changes GIRL CHILD into WOMAN able to bear children.

Note: As the level of OESTROGEN rises in the early stages of the cycle it inhibits (–) output of GnRH, FSH and LH.
About the 12th or 13th day the high level of oestrogen causes positive feedback (+). A sudden surge of LH (and of FSH) leads to OVULATION and the formation of the CORPUS LUTEUM. As the level of PROGESTERONE rises (along with OESTROGEN) it inhibits (–) GnRH, LH and FSH.
A few days before menstruation the CORPUS LUTEUM involutes. As the levels of OESTROGEN and PROGESTERONE fall, GnRH is freed from inhibition. Levels of FSH and LH begin to rise. The cycle starts again.

MENOPAUSE

Between the ages of 42 and 50 years OVARIAN tissue gradually ceases to respond to stimulation by <u>ANTERIOR PITUITARY GONADO-TROPHIC HORMONES.</u>

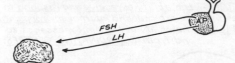

OVARIAN CYCLE becomes irregular and finally ceases ⟶ Ovary becomes small and fibrosed and no longer produces ripe Ova.

OESTROGEN and PROGESTERONE levels in Blood stream fall.

TISSUES of the body ⟋ begin to show changes which mark the end of REPRODUCTIVE LIFE.

Ducts⟍ Acini⟋ Atrophy

Sometimes final redistribution of fat → less typically feminine distribution.

Regression of <u>Secondary Sex Characteristics</u>.

Breasts shrink.

Hair becomes sparse in axillae and on pubis.

<u>Secondary Sex Organs</u> atrophy.

Fallopian tubes shrink.

Uterine Cycle and Menstruation cease. (Muscle and lining shrink).

Vaginal epithelium becomes thin.

External Genitalia shrink.

<u>Psychological and Personality</u> changes.

Decline in Sexual powers.

Emotional disturbances may occur – often accompanied by Vasomotor phenomena such as "Hot Flushes" (vasodilatation), excessive sweating and giddiness.

Uterus

Vagina

After the MENOPAUSE a woman is usually unable to bear children.

CHAPTER 8.

NERVOUS AND LOCOMOTOR SYSTEMS

The Nervous System is concerned with the INTEGRATION and CONTROL of all bodily functions.

It has specialized in IRRITABILITY — *the ability to receive and respond to messages from the external and internal environments*

and also in CONDUCTION — *the ability to transmit messages to and from CO-ORDINATING CENTRES.*

The NERVOUS SYSTEM
consists of
a
↓
CENTRAL PART —
The BRAIN and SPINAL CORD
↓
linked by an
outlying
or
↓
PERIPHERAL PART — *the Nerve Fibres*
to
TISSUES and ORGANS

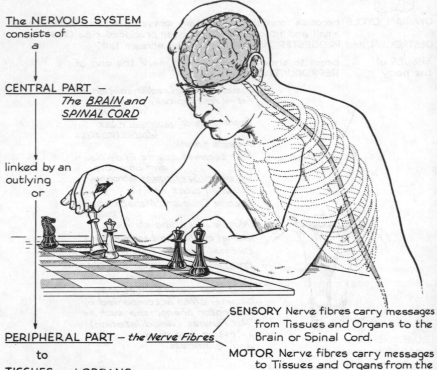

SENSORY Nerve fibres carry messages from Tissues and Organs to the Brain or Spinal Cord.

MOTOR Nerve fibres carry messages to Tissues and Organs from the Brain or Spinal Cord.

CEREBRUM

The largest part of the human brain is the CEREBRUM — made up of 2 CEREBRAL HEMISPHERES. Each of these is divided into LOBES.

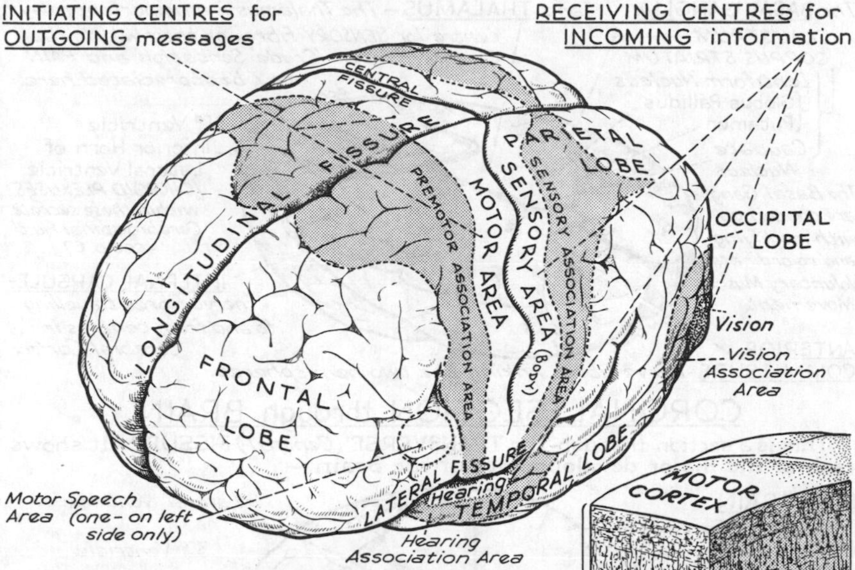

INITIATING CENTRES for
OUTGOING messages

RECEIVING CENTRES for
INCOMING information

CENTRAL FISSURE

CENTRAL FISSURE

LONGITUDINAL FISSURE

PARIETAL LOBE

PREMOTOR ASSOCIATION AREA

MOTOR AREA

SENSORY ASSOCIATION AREA (Body)

OCCIPITAL LOBE

Vision

FRONTAL LOBE

Vision Association Area

Motor Speech Area (one - on left side only)

LATERAL FISSURE

Hearing

TEMPORAL LOBE

Hearing Association Area

MOTOR CORTEX

GREY MATTER

BETZ CELL

WHITE MATTER

Large uncharted areas of the Cerebral Hemispheres are probably concerned with MENTAL PROCESSES such as *Intelligence, Memory, Judgement, Imagination, Creative* and *Conscious Thought.*

The surface of the brain shows many folds or CONVOLUTIONS. This has the effect of increasing the amount of GREY MATTER present. The GREY MATTER forms the outer layer or CORTEX. It contains the cell bodies of the NEURONES arranged in many interconnecting layers to form a 3-dimensional network.

About 90% of all Nerve Cells are in the Cerebral Cortex.

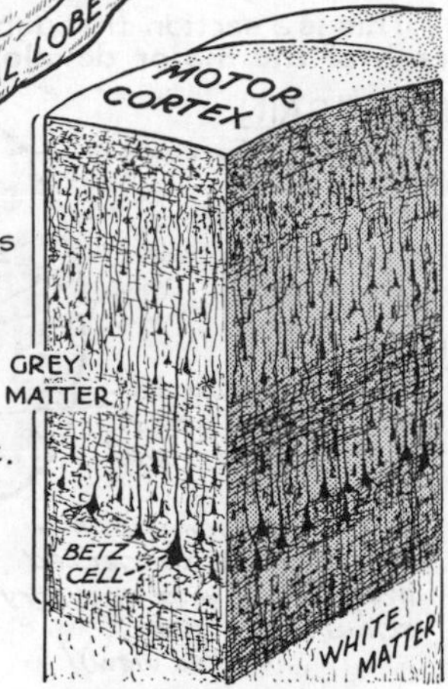

HORIZONTAL SECTION through BRAIN

This view shows surface GREY MATTER containing Nerve Cells and inner WHITE MATTER made up of Nerve Fibres. Deep in the substance of the Cerebral Hemispheres there are additional masses of GREY MATTER:-

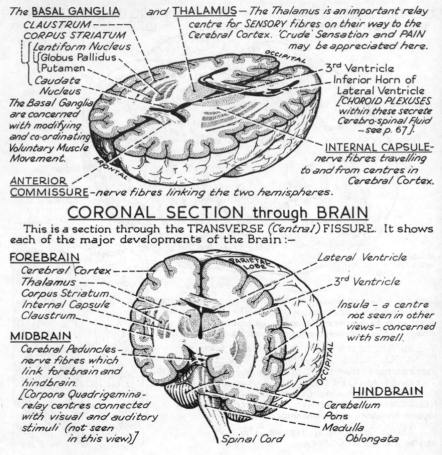

The **BASAL GANGLIA**
  CLAUSTRUM
  CORPUS STRIATUM
    Lentiform Nucleus
      {Globus Pallidus
      {Putamen
    Caudate
    Nucleus
The Basal Ganglia are concerned with modifying and co-ordinating Voluntary Muscle Movement.

and THALAMUS — *The Thalamus is an important relay centre for SENSORY fibres on their way to the Cerebral Cortex. 'Crude' Sensation and PAIN may be appreciated here.*

OCCIPITAL

3rd Ventricle
Inferior Horn of Lateral Ventricle
[CHOROID PLEXUSES within these secrete Cerebro-spinal Fluid — see p. 67].

FRONTAL

INTERNAL CAPSULE - *nerve fibres travelling to and from centres in Cerebral Cortex.*

ANTERIOR COMMISSURE - *nerve fibres linking the two hemispheres.*

CORONAL SECTION through BRAIN

This is a section through the TRANSVERSE (*Central*) FISSURE. It shows each of the major developments of the Brain :-

FOREBRAIN
  Cerebral Cortex
  Thalamus
  Corpus Striatum
  Internal Capsule
  Claustrum

MIDBRAIN
  Cerebral Peduncles- *nerve fibres which link forebrain and hindbrain.*
  [Corpora Quadrigemina- *relay centres connected with visual and auditory stimuli (not seen in this view)*]

PARIETAL LOBE

Lateral Ventricle

3rd Ventricle

Insula - a centre not seen in other views - concerned with smell.

OCCIPITAL

HINDBRAIN
  Cerebellum
  Pons
  Medulla
  Oblongata

Spinal Cord

VERTICAL SECTION through BRAIN

This is a Vertical Section through the LONGITUDINAL FISSURE — a deep cleft which separates the two Cerebral Hemispheres. At the bottom of this cleft are tracts of nerve fibres which link up the different LOBES of each hemisphere and also link the two hemispheres with each other — the CORPUS CALLOSUM.

FOREBRAIN

Cerebral Hemisphere

Thalamus—
-relay centres for sensation: pain appreciated here.

Hypothalamus
-contains centres for Autonomic Nervous System, e.g. Control of Heart, Blood pressure, Temperature, Metabolism, etc.

Opening of Lateral Ventricle

3RD Ventricle

Corpora Quadrigemina

Cerebellum
Centres concerned with balance and equilibrium. Important tracts link it with other parts of Brain and Spinal Cord.

Pituitary gland

PARIETAL LOBE
OCCIPITAL LOBE
CORPUS CALLOSUM
FORNIX
FRONTAL LOBE
PONS

MIDBRAIN
Receives impulses from Retina and Ear. Serves as a centre for Visual and Auditory Reflexes. In the Grey Matter are nerve cell bodies of III, IV Cranial nerves and the Red Nucleus which helps to control skilled muscular movements. The White Matter carries nerve fibres linking Red Nucleus with Cerebral Cortex, Thalamus, Cerebellum, Corpus Striatum and Spinal Cord. It also carries Ascending Sensory fibres in Lateral and Medial Lemnisci, and Descending Motor fibres on their way to Pons and Spinal Cord.

HINDBRAIN:
[PONS, CEREBELLUM, MEDULLA OBLONGATA]

Pons: Groups of Neurones form sensory nucleus of V and also nuclei of VI and VII Cranial nerves. Other nerve cells here relay impulses along their axons to Cerebellum and Cerebrum. Rubrospinal tract, Lateral and Medial Lemnisci pass through Pons and nerve fibres linking Cerebral Cortex with Medulla Oblongata and Spinal Cord.

Medulla Oblongata:
Groups of Neurones form Nuclei of VIII, IX, X, XI, XII Cranial nerves. Gracile and Cuneate Nuclei -second sensory neurones in cutaneous pathways. Tracts of Sensory fibres decussate and ascend to other side of Cerebral Cortex. Some fibres remain uncrossed. The larger part of each Motor pyramidal tract crosses and descends in other side of Spinal Cord.

CRANIAL NERVES

Twelve pairs of nerves arise directly from the undersurface of the Brain to supply Head and Neck and most of the viscera.

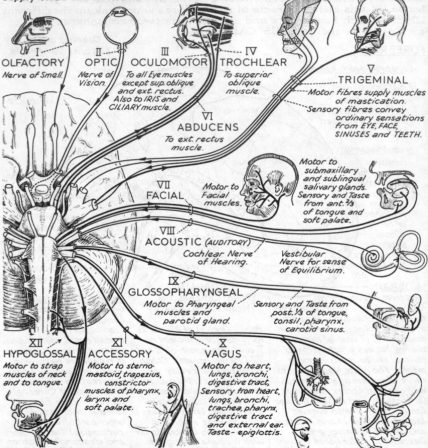

I
OLFACTORY
Nerve of Smell.

II
OPTIC
Nerve of Vision.

III
OCULOMOTOR
To all Eye muscles except sup. oblique and ext. rectus. Also to IRIS and CILIARY muscle.

IV
TROCHLEAR
To superior oblique muscle.

V
TRIGEMINAL
Motor fibres supply muscles of mastication. Sensory fibres convey ordinary sensations from EYE, FACE, SINUSES and TEETH.

VI
ABDUCENS
To ext. rectus muscle.

VII
FACIAL
Motor to Facial muscles.

Motor to submaxillary and sublingual salivary glands. Sensory and Taste from ant. ⅔ of tongue and soft palate.

VIII
ACOUSTIC *(AUDITORY)*
Cochlear Nerve of Hearing.

Vestibular Nerve for sense of Equilibrium.

IX
GLOSSOPHARYNGEAL
Motor to Pharyngeal muscles and parotid gland.

Sensory and Taste from post. ⅓ of tongue, tonsil, pharynx, carotid sinus.

XII
HYPOGLOSSAL
Motor to strap muscles of neck and to tongue.

XI
ACCESSORY
Motor to sterno-mastoid, trapezius, constrictor muscles of pharynx, larynx and soft palate.

X
VAGUS
Motor to heart, lungs, bronchi, digestive tract, Sensory from heart, lungs, bronchi, trachea, pharynx, digestive tract and external ear. Taste - epiglottis.

(After Frank H. NETTER, M.D., The Ciba Collection of Medical Illustrations)

120

SPINAL CORD

The SPINAL CORD lies within the Vertebral Canal. It is continuous above with the Medulla Oblongata.

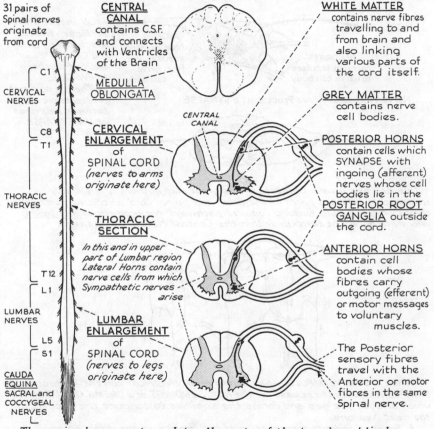

31 pairs of
Spinal nerves
originate
from cord

CERVICAL
NERVES

THORACIC
NERVES

LUMBAR
NERVES

CAUDA
EQUINA
SACRAL and
COCCYGEAL
NERVES

C1

C8
T1

T12
L1

L5
S1

CENTRAL CANAL
contains C.S.F.
and connects
with Ventricles
of the Brain

MEDULLA
OBLONGATA

CENTRAL
CANAL

CERVICAL ENLARGEMENT
of
SPINAL CORD
(nerves to arms
originate here)

THORACIC SECTION

In this and in upper
part of Lumbar region
Lateral Horns contain
nerve cells from which
Sympathetic nerves
arise

LUMBAR ENLARGEMENT
of
SPINAL CORD
(nerves to legs
originate here)

WHITE MATTER
contains nerve fibres
travelling to and
from brain and
also linking
various parts of
the cord itself.

GREY MATTER
contains nerve
cell bodies.

POSTERIOR HORNS
contain cells which
SYNAPSE with
ingoing (afferent)
nerves whose cell
bodies lie in the
POSTERIOR ROOT GANGLIA outside
the cord.

ANTERIOR HORNS
contain cell
bodies whose
fibres carry
outgoing (efferent)
or motor messages
to voluntary
muscles.

The Posterior
sensory fibres
travel with the
Anterior or motor
fibres in the same
Spinal nerve.

The spinal nerves travel to all parts of the trunk and limbs.

SYNAPSE

The structural unit of the Nervous System is the NEURONE. *(see pp. 12,13)*
Neurones are linked together in the Nervous System.....

..... Nerve Process to Nerve Cell body at a SYNAPSE

or

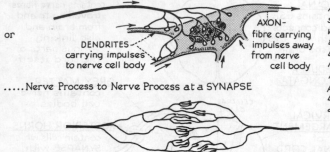

DENDRITES- carrying impulses to nerve cell body

AXON- fibre carrying impulses away from nerve cell body

AXON ends in small swellings – END FEET- which merely touch the DENDRITES or BODY of another NERVE CELL.
i.e. There is no direct protoplasmic union between neurones at the SYNAPSE.

..... Nerve Process to Nerve Process at a SYNAPSE

One neurone usually connects with a great many others often widely scattered in different parts of the Brain and Spinal Cord. In this way intricate chains of nerve cells forming complex pathways for <u>incoming</u> and <u>outgoing</u> information can be built up within the Central Nervous System.

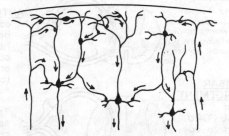

When the NERVE IMPULSE – like a very small electrical current — reaches a SYNAPSE it causes the release at the NERVE ENDINGS of a CHEMICAL SUBSTANCE which bridges the gap and forms the stimulus to conduct the Impulse to the next Neurone.
A Synapse permits transmission of the impulse in one direction only.

122

REFLEX ACTION

The NEURONE is the ANATOMICAL or STRUCTURAL UNIT of the Nervous System: the Nervous REFLEX is the PHYSIOLOGICAL or FUNCTIONAL UNIT.

A *Nervous Reflex* is an INVOLUNTARY ACTION caused by the STIMULATION of an AFFERENT (*sensory*) nerve ending or RECEPTOR.

The structural basis of Reflex Action is the REFLEX ARC. In its simplest form this consists of :-

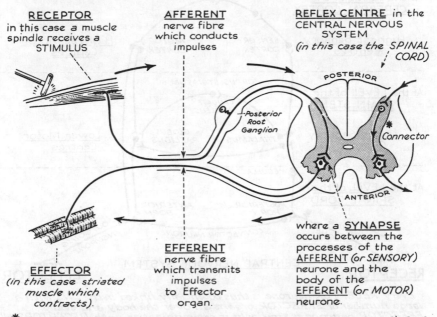

RECEPTOR
in this case a muscle spindle receives a STIMULUS

AFFERENT
nerve fibre which conducts impulses

REFLEX CENTRE in the CENTRAL NERVOUS SYSTEM
(*in this case the SPINAL CORD*)

POSTERIOR

Posterior Root Ganglion

Connector

ANTERIOR

EFFECTOR
(*in this case striated muscle which contracts*).

EFFERENT
nerve fibre which transmits impulses to Effector organ.

where a **SYNAPSE** occurs between the processes of the **AFFERENT** (*or SENSORY*) neurone and the body of the **EFFERENT** (*or MOTOR*) neurone.

* In most reflex arcs in man *afferent* and *efferent* neurones are linked by at least one *connector* neurone.

Reflexes form the basis of all Central Nervous System (C.N.S.) activity. They occur at all levels of the Brain and Spinal Cord. Important bodily functions such as movements of Respiration, Digestion, etc., are all controlled through Reflexes. We are made aware of some reflex acts: others occur without our knowledge.

"EDIFICE" of the C.N.S. (After R.C. GARRY)

In the majority of Reflex arcs in man a chain of many connector neurones is found. There may be link-ups with various levels of the Brain and Spinal Cord.

This diagram gives a simplified concept of the LINK-UP between different levels of the Central Nervous System.

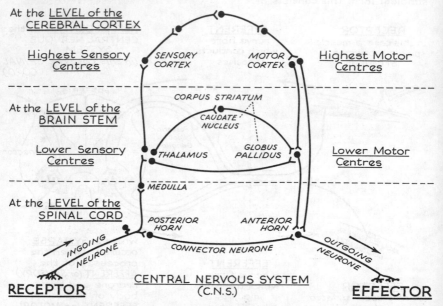

At the <u>LEVEL of the CEREBRAL CORTEX</u>

<u>Highest Sensory Centres</u>   SENSORY CORTEX   MOTOR CORTEX   <u>Highest Motor Centres</u>

CORPUS STRIATUM

At the <u>LEVEL of the BRAIN STEM</u>

CAUDATE NUCLEUS

<u>Lower Sensory Centres</u>   *THALAMUS*   *GLOBUS PALLIDUS*   <u>Lower Motor Centres</u>

MEDULLA

At the <u>LEVEL of the SPINAL CORD</u>

POSTERIOR HORN   ANTERIOR HORN

INGOING NEURONE   CONNECTOR NEURONE   OUTGOING NEURONE

RECEPTOR   <u>CENTRAL NERVOUS SYSTEM</u> (C.N.S.)   **EFFECTOR**

Every RECEPTOR neurone is thus potentially linked in the C.N.S with a large number of EFFECTOR organs all over the body and every EFFECTOR neurone is similarly in communication with RECEPTORS all over the body. Through such 'functional' link-ups, neurones in different parts of the Central Nervous System can influence each other. This makes it possible for 'CONDITIONED' REFLEXES to become established (for simple example see page 29). Such reflexes probably form the basis of all training so that it becomes difficult to say where REFLEX (or INVOLUNTARY) behaviour ends and purely VOLUNTARY behaviour begins.

REFLEX ACTION

Most REFLEX ACTIONS in man involve a great many REFLEX ARCS.

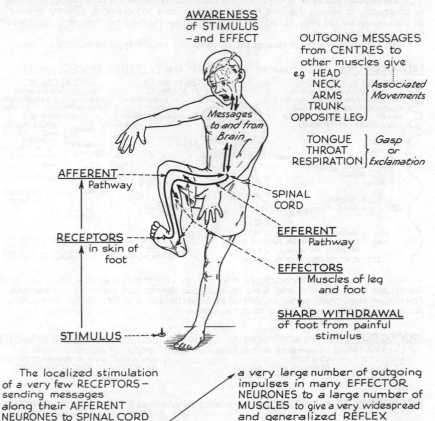

AWARENESS
of STIMULUS
—and EFFECT

OUTGOING MESSAGES
from CENTRES to
other muscles give
e.g. HEAD
NECK
ARMS   } Associated
TRUNK   Movements
OPPOSITE LEG

TONGUE   } Gasp
THROAT   or
RESPIRATION } Exclamation

Messages
to and from
Brain

AFFERENT
Pathway

RECEPTORS
in skin of
foot

STIMULUS

SPINAL
CORD

EFFERENT
Pathway

EFFECTORS
Muscles of leg
and foot

SHARP WITHDRAWAL
of foot from painful
stimulus

The localized stimulation of a very few RECEPTORS — sending messages along their AFFERENT NEURONES to SPINAL CORD and to BRAIN — has led to → a very large number of outgoing impulses in many EFFECTOR NEURONES to a large number of MUSCLES to give a very widespread and generalized REFLEX RESPONSE.

This is possible because each receptor neurone is potentially connected within the Central Nervous System with all effector neurones.

9

SENSE ORGANS

Man's awareness of the world is limited to those forms of energy, physical or chemical, to which he has Receptors designed to respond. (Many "events" in the universe go unnoticed by man because he has no Sense Organ which can respond to them.)

Each Sense Organ is designed to respond to one type of stimulation.

EXTEROCEPTORS are stimulated by events in the <u>EXTERNAL ENVIRONMENT</u>.

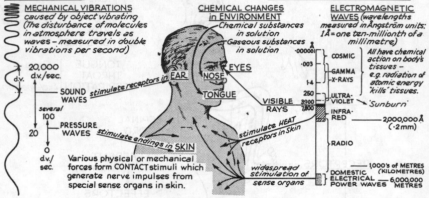

MECHANICAL VIBRATIONS *caused by object vibrating (The disturbance of molecules in atmosphere travels as waves – measured in double vibrations per second)*

20,000 d.v./sec.

SOUND WAVES *stimulate receptors in* EAR

several 100

20 PRESSURE WAVES *stimulate endings in* SKIN

0 d.v./sec. Various physical or mechanical forces form CONTACT stimuli which generate nerve impulses from special sense organs in skin.

CHEMICAL CHANGES in ENVIRONMENT
—Chemical substances in solution
—Gaseous substances in solution

EYES
NOSE
TONGUE

stimulate HEAT *receptors in Skin*

widespread stimulation of sense organs

ELECTROMAGNETIC WAVES *(wavelengths measured in Ångström units: 1Å = one ten-millionth of a millimetre)*

·00001	COSMIC	All have chemical action on body's tissues – e.g. radiation of atomic energy 'kills' tissues.
·003	GAMMA
1-4	X-RAYS
250	ULTRA-VIOLET	'Sunburn'
3900	VISIBLE RAYS
7,800	INFRA-RED	2,000,000 Å (·2mm)
	RADIO
		1,000's of METRES (KILOMETRES)
	DOMESTIC ELECTRICAL POWER WAVES	6,000,000 METRES

Exteroceptors may convey information to CONSCIOUSNESS with AWARENESS or SENSATION and lead to suitable RESPONSES planned in CEREBRAL CORTEX or they may serve as AFFERENT pathways for REFLEX (or INVOLUNTARY) ACTION with or without rising to consciousness.

PROPRIOCEPTORS are stimulated by changes in <u>LOCOMOTOR SYSTEM</u> of body

LABYRINTH	*movements and position of head*	SENSE of EQUILIBRIUM or
MUSCLES	*stretch*	BALANCE and AWARENESS
TENDONS	*tension and stretch*	of POSITION and MOVEMENT
JOINTS	*stretch and pressure*	of BODY in space.

INTEROCEPTORS in <u>VISCERA</u> are stimulated by changes in <u>INTERNAL ENVIRONMENT</u> *(e.g. by distension in hollow organs).*

Much of the PROPRIO- and INTEROCEPTOR information never rises to consciousness.

<u>*Overstimulation*</u> *of <u>any receptor</u> can give rise to sensation of <u>PAIN</u>.*

Most receptors show <u>ADAPTATION</u> – if continuously stimulated they send reduced numbers of impulses to the brain.

126

SMELL

Smell is a CHEMICAL SENSE, i.e. the receptors respond to CHEMICAL STIMULI.
To arouse the sensation a substance must first be in a GASEOUS STATE
then go into SOLUTION.

<u>The ORGAN of SMELL is the NOSE</u> — *Also serves as the main air passage to Respiratory System.*

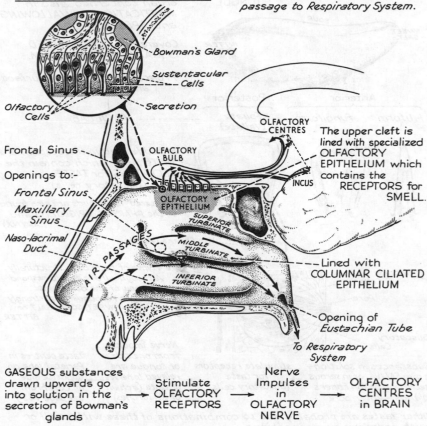

Bowman's Gland

Sustentacular Cells

Secretion

Olfactory Cells

OLFACTORY CENTRES

The upper cleft is lined with specialized OLFACTORY EPITHELIUM which contains the RECEPTORS for SMELL.

Frontal Sinus

OLFACTORY BULB

INCUS

Openings to:-

Frontal Sinus

OLFACTORY EPITHELIUM

Maxillary Sinus

SUPERIOR TURBINATE

Naso-lacrimal Duct

MIDDLE TURBINATE

AIR PASSAGES

INFERIOR TURBINATE

Lined with COLUMNAR CILIATED EPITHELIUM

Opening of *Eustachian Tube*

To Respiratory System

GASEOUS substances drawn upwards go into solution in the secretion of Bowman's glands → Stimulate OLFACTORY RECEPTORS → Nerve Impulses in OLFACTORY NERVE → OLFACTORY CENTRES in BRAIN

TASTE

Taste is a CHEMICAL SENSE, i.e. receptors respond to CHEMICAL STIMULI.
To arouse the sensation a substance must be in SOLUTION.

The essential <u>ORGAN of TASTE</u> is the _____ TONGUE

The VOLUNTARY muscular organ concerned also in MASTICATION, SWALLOWING and SPEECH.

Covered with STRATIFIED SQUAMOUS EPITHELIUM.

Projections on its upper surface are called

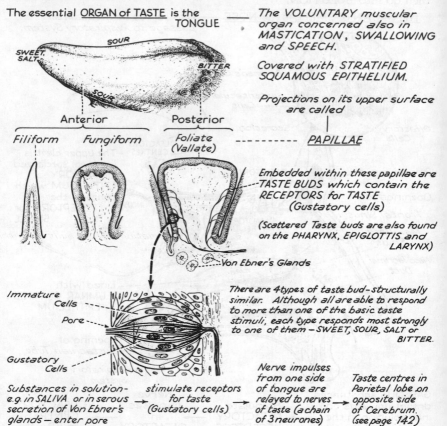

SOUR

SWEET. SALT.

BITTER

SOUR

Anterior — Posterior

Filiform — Fungiform — Foliate (Vallate) ---------- *PAPILLAE*

Embedded within these papillae are TASTE BUDS which contain the RECEPTORS for TASTE (Gustatory cells)

(Scattered Taste buds are also found on the PHARYNX, EPIGLOTTIS and LARYNX)

Von Ebner's Glands

Immature Cells

Pore

Gustatory Cells

There are 4 types of taste bud—structurally similar. Although all are able to respond to more than one of the basic taste stimuli, each type responds most strongly to one of them — SWEET, SOUR, SALT or BITTER.

Substances in solution—e.g. in SALIVA or in serous secretion of Von Ebner's glands — enter pore → stimulate receptors for taste (Gustatory cells) → Nerve impulses from one side of tongue are relayed to nerves of taste (a chain of 3 neurones) → Taste centres in Parietal lobe on opposite side of Cerebrum. (see page 142)

Other tastes are probably due to combinations of these with smell or with ordinary skin sensations.

EYE

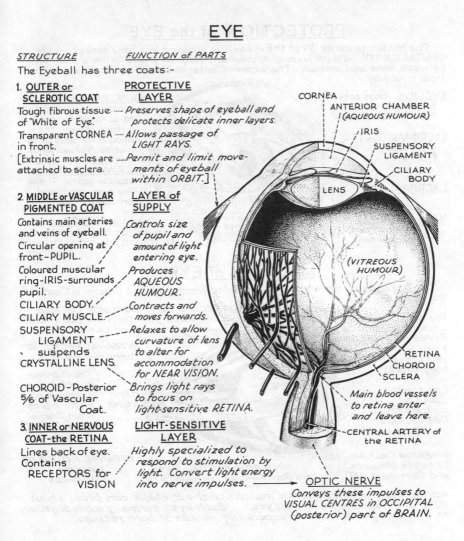

STRUCTURE

FUNCTION of PARTS

The Eyeball has three coats:-

1. OUTER or SCLEROTIC COAT

PROTECTIVE LAYER

Tough fibrous tissue of "White of Eye". ···· *Preserves shape of eyeball and protects delicate inner layers.*

Transparent CORNEA in front. ···· *Allows passage of LIGHT RAYS.*

[Extrinsic muscles are attached to sclera. ···· *Permit and limit movements of eyeball within ORBIT.]*

2. MIDDLE or VASCULAR PIGMENTED COAT

LAYER of SUPPLY

Contains main arteries and veins of eyeball.

Circular opening at front–PUPIL. ··· *Controls size of pupil and amount of light entering eye.*

Coloured muscular ring–IRIS–surrounds pupil. ··· *Produces AQUEOUS HUMOUR.*

CILIARY BODY.

CILIARY MUSCLE. ··· *Contracts and moves forwards.*

SUSPENSORY LIGAMENT suspends CRYSTALLINE LENS. ··· *Relaxes to allow curvature of lens to alter for accommodation for NEAR VISION.*

CHOROID–Posterior 5/6 of Vascular Coat. ··· *Brings light rays to focus on light-sensitive RETINA.*

3. INNER or NERVOUS COAT-the RETINA

LIGHT-SENSITIVE LAYER

Lines back of eye. Contains RECEPTORS for VISION ··· *Highly specialized to respond to stimulation by light. Convert light energy into nerve impulses.*

CORNEA
ANTERIOR CHAMBER (AQUEOUS HUMOUR)
IRIS
SUSPENSORY LIGAMENT
CILIARY BODY
LENS
(VITREOUS HUMOUR)
RETINA
CHOROID
SCLERA
Main blood vessels to retina enter and leave here.
CENTRAL ARTERY of the RETINA

⟶ OPTIC NERVE
Conveys these impulses to VISUAL CENTRES in OCCIPITAL (posterior) part of BRAIN.

PROTECTION of the EYE

The hidden posterior 4/5 of the eyeball is encased in a bony socket — the ORBITAL CAVITY. A thick layer of *Areolar* and *Adipose* tissue forms a cushion between bone and eyeball. The exposed anterior 1/5 of the eyeball is protected from injury by:-

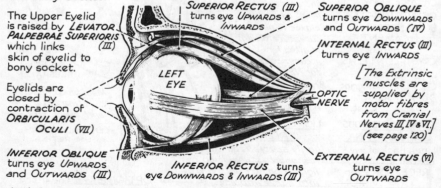

EYELIDS – close reflexly to protect eye from foreign bodies

LACRIMAL DUCTS. drain tears from eye
↓
LACRIMAL SAC
↓
NASO-LACRIMAL DUCT drains tears into back of nose

ORBICULARIS OCULI MUSCLE

CONJUNCTIVA – a delicate membrane lining eyelids and covering exposed surface of eye. Its smooth surfaces glide over each other when lids open and close.

LACRIMAL GLANDS – secrete tears. These flow over, wash and lubricate surface of eye.

Openings of TARSAL GLANDS – secrete a fluid to prevent lids from sticking together.

MUSCLES of EYE

The Eyeballs are moved by Small Muscles which link the Sclerotic Coat to the Bony Socket.

The Upper Eyelid is raised by *LEVATOR PALPEBRAE SUPERIORIS* (III) which links skin of eyelid to bony socket.

Eyelids are closed by contraction of *ORBICULARIS OCULI* (VII)

INFERIOR OBLIQUE turns eye *UPWARDS* and *OUTWARDS* (III)

LEFT EYE

SUPERIOR RECTUS (III) turns eye *UPWARDS & INWARDS*

INFERIOR RECTUS turns eye *DOWNWARDS & INWARDS* (III)

SUPERIOR OBLIQUE turns eye *DOWNWARDS* and *OUTWARDS* (IV)

INTERNAL RECTUS (III) turns eye *INWARDS*

[The Extrinsic muscles are supplied by motor fibres from Cranial Nerves III, IV & VI.] (see page 120)

OPTIC NERVE

EXTERNAL RECTUS (VI) turns eye *OUTWARDS*

Acting together, the Extrinsic muscles of the Eyeballs can bring about ROTATORY movements of the Eyes. Both eyes normally move together so that images fall on corresponding points of both retinae.

ACTION of LENS

The normal lens brings light rays to a sharp focus upside down on the retina. It can do this whether we are looking at an object far away or one close at hand. The curvature increases reflexly to accommodate for near vision.

Rays of light coming from every point of a DISTANT object (over 20 feet away) are PARALLEL.

They pass through the CORNEA, AQUEOUS HUMOUR

and LENS which refract them to a sharp focus — upside down and reversed from side to side — on the Retina.

The conscious mind learns to interpret the image and project it to its true position in space.

Rays of light coming from a NEAR object (less than 20 feet away) RADIATE from every point

A more convex lens is required to bring these rays to a sharp focus on the Retina.
Ciliary muscle contracts →
Suspensory lig^t slackens →
tension on capsule of lens slackens → lens bulges.

If the EYEBALL is <u>too short</u>, rays from a distant object are brought into focus BEHIND the Retina when the ciliary muscle is relaxed.

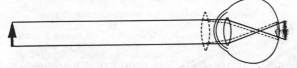

Vision is blurred.

This is longsightedness or HYPERMETROPIA.

The longsighted eye has to accommodate even for distant vision: i.e. Ciliary muscles contract to give a more convex lens and distant objects are then seen clearly. This limits amount of accommodating power left for near objects and the nearest point for sharp vision is then further away. It can be corrected by fitting spectacles with an additional convex lens.

If the EYEBALL is <u>too long</u>, rays from a distant object are brought into focus IN FRONT of the Retina.

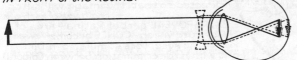

This is shortsightedness or MYOPIA — only objects near the eye can be seen clearly.

It can be corrected by using a concave lens.

RETINA

Sections of the Retina examined under the microscope show 8 layers:—

LAYER of PIGMENT CELLS next to Choroid Coat.

In <u>bright</u> light pigment granules migrate into the cell processes lying between rods and cones. This prevents spread of light from one receptor to another. In <u>dim</u> light the granules are confined to the cell body. When light strikes RODS and CONES, impulses are set up which are transmitted thus:—

RODS and CONES → INTERMEDIATE NEURONES → GANGLION CELLS or INTEGRATING NEURONES → Nerve fibres converge on Optic Papilla to leave as the **OPTIC NERVE**

↓

to Visual Area of **CEREBRAL CORTEX**

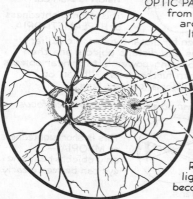

FOVEA CENTRALIS

Note:— LIGHT rays must pass through all these layers except the pigment cell layer to reach and stimulate Receptors.

The <u>FUNDUS OCULI</u> is the part of the Retina seen through the Ophthalmoscope.

OPTIC PAPILLA - or *'BLIND SPOT'* - The nerve fibres from all parts of the retina converge on this area to leave the eyeball as the OPTIC NERVE. It has no RODS or CONES and therefore is not itself sensitive to light.

RETINAL BLOOD VESSELS enter or leave the eyeball here.

MACULA LUTEA-or *'YELLOW SPOT'*- with **FOVEA CENTRALIS** -area of acute vision- contains CONES only (the Receptors stimulated in Bright and Coloured light). When we look at an object the eyes are directed so that the image will fall on the fovea of each eye.

EXTRAFOVEAL part of Retina - area of less acute vision- CONES become fewer - RODS (the Receptors stimulated in Dim light with no discrimination between Colours) become more numerous towards the peripheral part of the Retina.

132

MECHANISM of VISION

WHITE LIGHT is really due to the fusion of *coloured lights*. These coloured lights are separated by shining a beam of white light through a glass prism. This is called the VISIBLE SPECTRUM.

7,800Å	7,000Å	6,000Å				5,000Å			3,900Å
RED		ORANGE	YELLOW	YELL.-GR.	GREEN	BLUE-GREEN	BLUE	VIOLET	

If light is bright or intense the spectrum appears brightest to man's eye in the orange band (6,100 Å).

If brightness or intensity of light source is gradually reduced, colour perception is gradually lost and the spectrum appears as a luminous band with a very dark area in the red band and with its brightest part in the green band (5,300Å).

Visual Receptors contain PIGMENTS which break down chemically in the presence of specific wavelengths of light. This forms a chemical stimulus which in some way triggers off nerve impulses which travel from the retina to the cerebral cortex.

SCOTOPIC VISION is vision in DIM LIGHT. It depends on the RODS

RODS are of one type ————and give————MONOCHROMATIC VISION

In DIM LIGHT

RODS contain RHODOPSIN ("VISUAL
or   PURPLE")
SCOTOPSIN

$\xrightarrow[\text{Pigment gradually splits}]{\text{absorbs light}}$ RETINENE + OPSIN
(a protein)

↓

*Chemical Stimulus
triggers off
Nerve Impulse*

[In BRIGHT LIGHT

RHODOPSIN- very rapid breakdown with "bleaching."]

In DARKNESS

RHODOPSIN $\xleftarrow{\text{regeneration}}$ RETINENE + OPSIN
(regenerated from Vit.A)

When a person passes from a brightly lit scene to darkness he is temporarily blinded. After about ½ hour he sees well. This adjustment or increase in sensitivity is called DARK ADAPTATION.

*As light brightness or intensity increases rods lose their sensitivity
and cease to respond.*

MECHANISM of VISION

PHOTOPIC VISION is vision in BRIGHT LIGHT. It depends on the CONES

CONES are thought to be of 3 types —— giving TRICHROMATIC VISION.
Each type with a
different
photosensitive
VISUAL
PIGMENT with its own wavelength to which it is sensitive, which

Note:- The Photosensitive pigments in cones have not yet been isolated.

it absorbs and by which it is broken down to form the
chemical stimulus.

"RED" receptors absorb YELLOW—ORANGE light
"GREEN" receptors absorb GREEN light
"BLUE" receptors absorb BLUE light

All 3 types of "PHOTOPSINS" are probably stimulated in roughly equal
proportions when WHITE light falls on retina:
2 or more in varying proportions when OTHER COLOURS of light fall on
retina.

The various types of colour blindness could be explained in terms of the
absence or deficiency of one or more of these special receptors.

[*A colour sensation has 3 qualities:-*
   HUE ——————— depends largely on wavelength.
   SATURATION — purity —
      A "saturated" colour has no white light mixed with it.
      An "unsaturated" colour has some white light mixed with it.
   INTENSITY —— brightness — depends largely on "strength" of the light.]

As the intensity of light is reduced the Cones cease to respond and
the Rods take over.

When a person passes from darkness to bright light he is dazzled
but after a short time he sees well again.
   This adjustment or decrease in sensitivity on exposure to bright light
is called LIGHT ADAPTATION.

VISUAL PATHWAYS to the BRAIN

The RECEPTORS for VISION are linked by a chain of Neurones with RECEIVING and INTEGRATING CENTRES in the OCCIPITAL LOBES of the CEREBRAL CORTEX.

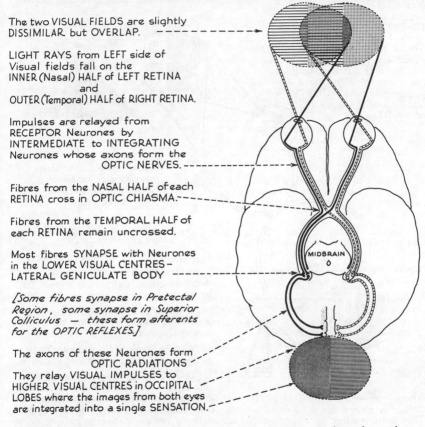

The two VISUAL FIELDS are slightly DISSIMILAR but OVERLAP.

LIGHT RAYS from LEFT side of Visual fields fall on the INNER (Nasal) HALF of LEFT RETINA and OUTER (Temporal) HALF of RIGHT RETINA.

Impulses are relayed from RECEPTOR Neurones by INTERMEDIATE to INTEGRATING Neurones whose axons form the OPTIC NERVES.

Fibres from the NASAL HALF of each RETINA cross in OPTIC CHIASMA.

Fibres from the TEMPORAL HALF of each RETINA remain uncrossed.

Most fibres SYNAPSE with Neurones in the LOWER VISUAL CENTRES – LATERAL GENICULATE BODY

[Some fibres synapse in Pretectal Region, some synapse in Superior Colliculus — these form afferents for the OPTIC REFLEXES.]

The axons of these Neurones form OPTIC RADIATIONS
They relay VISUAL IMPULSES to HIGHER VISUAL CENTRES in OCCIPITAL LOBES where the images from both eyes are integrated into a single SENSATION.

MIDBRAIN

Note:- One side of the OCCIPITAL CORTEX receives impressions from the FIELD of VISION on the opposite side.

STEREOSCOPIC VISION

When we look at some object or scene the view seen by the RIGHT EYE is slightly different from the view seen by the LEFT EYE.

These two DISSIMILAR RETINAL IMAGES are fused in the Visual Centres of the Brain to give a 3-dimensional picture – an appreciation of DEPTH as well as of HEIGHT and WIDTH.

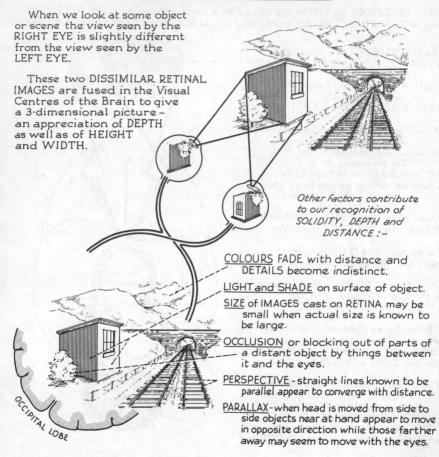

Other factors contribute to our recognition of SOLIDITY, DEPTH and DISTANCE :-

<u>COLOURS</u> FADE with distance and DETAILS become indistinct.

<u>LIGHT</u> and <u>SHADE</u> on surface of object.

<u>SIZE</u> of IMAGES cast on RETINA may be small when actual size is known to be large.

<u>OCCLUSION</u> or blocking out of parts of a distant object by things between it and the eyes.

<u>PERSPECTIVE</u> - straight lines known to be parallel appear to converge with distance.

<u>PARALLAX</u> - when head is moved from side to side objects near at hand appear to move in opposite direction while those farther away may seem to move with the eyes.

OCCIPITAL LOBE

By complex mental processes these points are interpreted in terms of Distance and Depth.

EAR

The ear has 3 separate parts, each with different rôles in the mechanism of HEARING:-

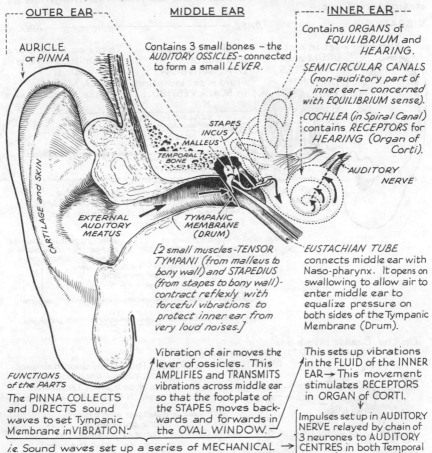

--- OUTER EAR ---

AURICLE or *PINNA*

CARTILAGE and SKIN

EXTERNAL AUDITORY MEATUS

MIDDLE EAR

Contains 3 small bones – the *AUDITORY OSSICLES* - connected to form a small *LEVER*.

STAPES
INCUS
MALLEUS
TEMPORAL BONE

TYMPANIC MEMBRANE (DRUM)

[2 small muscles - TENSOR TYMPANI (from malleus to bony wall) and STAPEDIUS (from stapes to bony wall) - contract reflexly with forceful vibrations to protect inner ear from very loud noises.]

----- INNER EAR ---

Contains *ORGANS* of *EQUILIBRIUM* and *HEARING*.

SEMICIRCULAR CANALS (non-auditory part of inner ear — concerned with *EQUILIBRIUM* sense).

COCHLEA (in Spiral Canal) contains *RECEPTORS* for *HEARING* (Organ of Corti).

AUDITORY NERVE

EUSTACHIAN TUBE connects middle ear with Naso-pharynx. It opens on swallowing to allow air to enter middle ear to equalize pressure on both sides of the Tympanic Membrane (Drum).

FUNCTIONS of the PARTS

The PINNA COLLECTS and DIRECTS sound waves to set Tympanic Membrane in VIBRATION.

Vibration of air moves the lever of ossicles. This AMPLIFIES and TRANSMITS vibrations across middle ear so that the footplate of the STAPES moves backwards and forwards in the OVAL WINDOW. —

This sets up vibrations in the FLUID of the INNER EAR → This movement stimulates RECEPTORS in ORGAN of CORTI.

↓

Impulses set up in AUDITORY NERVE relayed by chain of 3 neurones to AUDITORY CENTRES in both Temporal lobes (*see page 117*)

i.e. Sound waves set up a series of MECHANICAL STIMULI →

MECHANISM of HEARING

This is most readily understood if the COCHLEA is imagined as straightened out:—

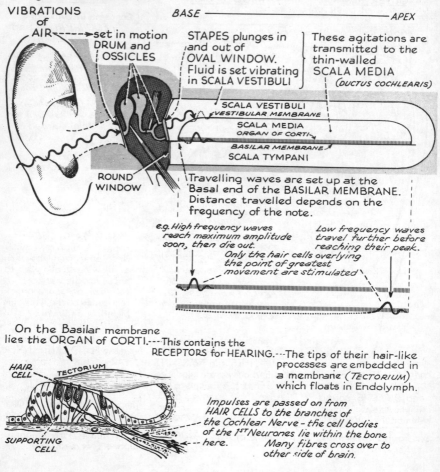

VIBRATIONS of AIR ----→ set in motion DRUM and OSSICLES

BASE ———————————————— APEX

STAPES plunges in and out of OVAL WINDOW. Fluid is set vibrating in SCALA VESTIBULI

These agitations are transmitted to the thin-walled SCALA MEDIA *(DUCTUS COCHLEARIS)*

SCALA VESTIBULI
VESTIBULAR MEMBRANE
SCALA MEDIA
ORGAN OF CORTI
BASILAR MEMBRANE
SCALA TYMPANI

ROUND WINDOW

Travelling waves are set up at the Basal end of the BASILAR MEMBRANE. Distance travelled depends on the frequency of the note.

e.g. High frequency waves reach maximum amplitude soon, then die out.

Low frequency waves travel further before reaching their peak.

Only the hair cells overlying the point of greatest movement are stimulated

On the Basilar membrane lies the ORGAN of CORTI.---This contains the RECEPTORS for HEARING.---The tips of their hair-like processes are embedded in a membrane (*TECTORIUM*) which floats in Endolymph.

HAIR CELL   TECTORIUM

SUPPORTING CELL

Impulses are passed on from HAIR CELLS to the branches of the Cochlear Nerve - the cell bodies of the 1ST Neurones lie within the bone here. Many fibres cross over to other side of brain.

SPECIAL PROPRIOCEPTORS

The "Special" Proprioceptors of the body are found in the Non-Auditory part of the Inner Ear — the Labyrinth. They are stimulated by movements or change of position of the head in space.

Three **SEMICIRCULAR CANALS** — *in each inner ear — one in each of three planes of space* - contain Receptors.

Membranous tubes containing Endolymph

These receptors, situated in the Ampulla of each canal, are stimulated mechanically by the *starting or stopping of rotatory movements* of the head in space.

One end of each canal has a swelling - the AMPULLA

Both ends of each canal open into the UTRICLE

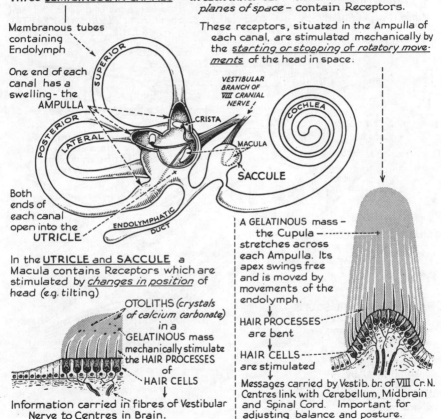

SUPERIOR

POSTERIOR

LATERAL

CRISTA

VESTIBULAR BRANCH OF VIII CRANIAL NERVE

COCHLEA

MACULA

SACCULE

ENDOLYMPHATIC DUCT

In the **UTRICLE** and **SACCULE** a Macula contains Receptors which are stimulated by *changes in position* of head (e.g. tilting)

OTOLITHS (*crystals of calcium carbonate*) in a GELATINOUS mass mechanically stimulate the HAIR PROCESSES of HAIR CELLS

Information carried in fibres of Vestibular Nerve to Centres in Brain.

A GELATINOUS mass — the Cupula — stretches across each Ampulla. Its apex swings free and is moved by movements of the endolymph.

HAIR PROCESSES are bent

HAIR CELLS are stimulated

Messages carried by Vestib. br. of VIII Cr. N. Centres link with Cerebellum, Midbrain and Spinal Cord. Important for adjusting balance and posture.

GENERAL PROPRIOCEPTORS

Proprioceptors are the sense organs stimulated by MOVEMENT of the body itself. They make us aware of the movement or position of the body in space and of the various parts of the body to each other. They are important as ingoing afferent pathways in reflexes for adjusting Posture and Tone. They may be linked with centres in (a) CEREBELLUM or (b) PARIETAL LOBE of opposite side of brain where awareness of muscle and joint sense is appreciated.   General Proprioceptors are found in SKELETAL MUSCLES, TENDONS and JOINTS.

GOLGI ORGAN — in tendons —
stimulated by tension which occurs
when muscle is STRETCHED and
when it is CONTRACTED

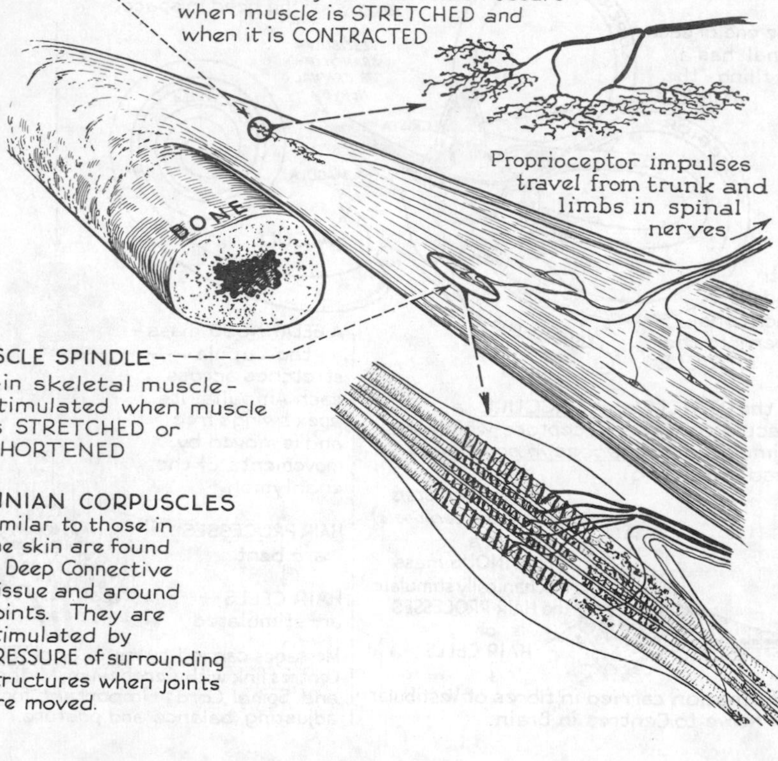

Proprioceptor impulses
travel from trunk and
limbs in spinal
nerves

BONE

MUSCLE SPINDLE - - - - - - -
—in skeletal muscle—
stimulated when muscle
is STRETCHED or
SHORTENED

PACINIAN CORPUSCLES
similar to those in
the skin are found
in Deep Connective
Tissue and around
Joints. They are
stimulated by
PRESSURE of surrounding
structures when joints
are moved.

CUTANEOUS SENSATION

There are 5 basic skin sensations — TOUCH, PRESSURE, PAIN, WARMTH and COLD. There is much controversy as to how these are registered. In some areas they appear to be served by special nerve endings (sensory receptors or end-organs) in the SKIN. These end-organs are not uniformly distributed over the whole body surface. *(Eg. Touch "endings" are very numerous in hands and feet but are much less frequent in the skin of the back.)*

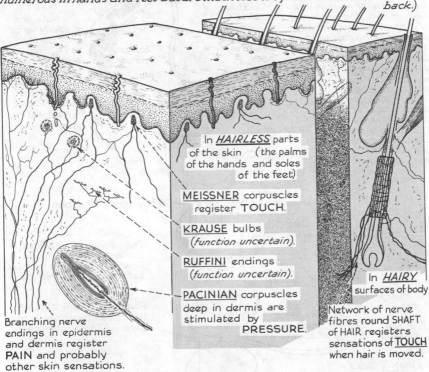

In *HAIRLESS* parts of the skin (the palms of the hands and soles of the feet)

<u>MEISSNER</u> corpuscles register **TOUCH**.

<u>KRAUSE</u> bulbs *(function uncertain).*

<u>RUFFINI</u> endings *(function uncertain).*

<u>PACINIAN</u> corpuscles deep in dermis are stimulated by **PRESSURE**.

Branching nerve endings in epidermis and dermis register **PAIN** and probably other skin sensations.

In *HAIRY* surfaces of body

Network of nerve fibres round SHAFT of HAIR registers sensations of <u>TOUCH</u> when hair is moved.

Tickling, itching, softness, hardness, wetness are probably due to stimulation of two or more of these special endings and to a blending of the sensations in the Brain.

10

SENSORY PATHWAYS and CORTEX

The nerve endings registering Pain, Warmth or Cold, Touch and Pressure, are linked by a chain of 3 Neurones with final RECEIVING CENTRES in the SENSORY area of the PARIETAL LOBES. The exact points to which impulses come from different regions are indicated.

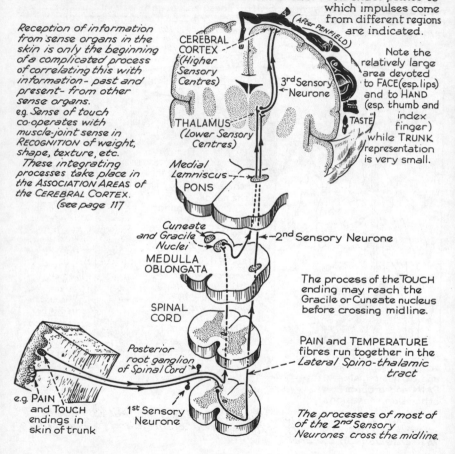

Reception of information from sense organs in the skin is only the beginning of a complicated process of correlating this with information- past and present- from other sense organs.
e.g. Sense of touch co-operates with muscle-joint sense in RECOGNITION of weight, shape, texture, etc.
These integrating processes take place in the ASSOCIATION AREAS of the CEREBRAL CORTEX.
(see page 117

CEREBRAL CORTEX (Higher Sensory Centres)

(After PENFIELD)

3rd Sensory Neurone

Note the relatively large area devoted to FACE (esp. lips) and to HAND (esp. thumb and index finger) while TRUNK representation is very small.

TASTE

THALAMUS (Lower Sensory Centres)

Medial Lemniscus

PONS

Cuneate and Gracile Nuclei

2nd Sensory Neurone

MEDULLA OBLONGATA

SPINAL CORD

The process of the TOUCH ending may reach the Gracile or Cuneate nucleus before crossing midline.

Posterior root ganglion of Spinal Cord

PAIN and TEMPERATURE fibres run together in the *Lateral Spino-thalamic tract*

e.g. PAIN and TOUCH endings in skin of trunk

1st Sensory Neurone

The processes of most of of the 2nd Sensory Neurones cross the midline.

MOTOR PATHWAYS to TRUNK and LIMBS

The CONTROLLING CENTRES in the MOTOR CORTEX are linked by 2 Neurones with the EFFECTOR ORGANS – the VOLUNTARY MUSCLES.

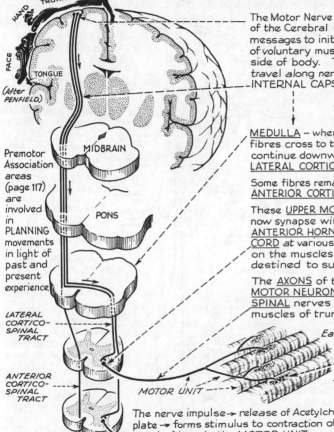

(After PENFIELD)

Premotor Association areas (page 117) are involved in PLANNING movements in light of past and present experience

LATERAL CORTICO-SPINAL TRACT

ANTERIOR CORTICO-SPINAL TRACT

The Motor Nerve Cells on one side of the Cerebral Cortex send out messages to initiate movements of voluntary muscles on opposite side of body. These messages travel along nerve fibres in the INTERNAL CAPSULE.

MEDULLA – where most of the fibres cross to the other side and continue downwards in the LATERAL CORTICO-SPINAL TRACT.

Some fibres remain uncrossed in ANTERIOR CORTICO-SPINAL TRACT.

These UPPER MOTOR NEURONES now synapse with cells in the ANTERIOR HORN of the SPINAL CORD at various levels depending on the muscles they are destined to supply.

The AXONS of these LOWER MOTOR NEURONES travel in the SPINAL nerves to the skeletal muscles of trunk and limbs.

Each branch releases Acetylcholine at a motor end-plate in a single muscle fibre.

MOTOR UNIT

The nerve impulse → release of Acetylcholine at motor end-plate → forms stimulus to contraction of the group of muscle fibres in the MOTOR UNIT.

CONTROL of MUSCLE MOVEMENT

The Cerebral Cortex initiates purposeful movements. During such movements proprioceptors are continually supplying information to Cerebellum about the changing positions of muscles and joints. The Cerebellum is then responsible for collating this information and sending out impulses to bring about the smooth co-ordinated movements of different groups of muscles.

This involves co-operation of other centres in the
EXTRAPYRAMIDAL SYSTEM

CORPUS STRIATUM

CAUDATE NUCLEUS
LENTIFORM NUCLEUS:
    PUTAMEN,
    GLOBUS
    PALLIDUS

These centres have important 2-way connections with each other and with HIGHER and LOWER MOTOR and SENSORY Centres.

CEREBELLUM

MIDBRAIN and PONS
    RED NUCLEUS
    SUBSTANTIA NIGRA
    RETICULAR NUCLEI
    (scattered groups of neurones)
    TECTUM
    VESTIBULAR NUCLEI

Where part of the Cerebellum or Extrapyramidal System is destroyed by disease varying types of RIGIDITY, TREMOR and UNCO-ORDINATED MUSCLE MOVEMENT result.

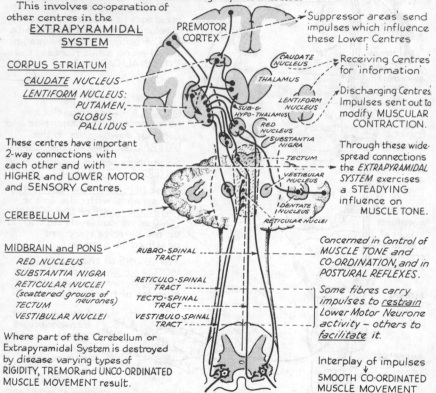

PREMOTOR CORTEX

CAUDATE NUCLEUS
THALAMUS
LENTIFORM NUCLEUS
SUB- & HYPO-THALAMUS
RED NUCLEUS
SUBSTANTIA NIGRA
TECTUM
VESTIBULAR NUCLEUS
DENTATE NUCLEUS
RETICULAR NUCLEI

RUBRO-SPINAL TRACT
RETICULO-SPINAL TRACT
TECTO-SPINAL TRACT
VESTIBULO-SPINAL TRACT

'Suppressor areas' send impulses which influence these Lower Centres

'Receiving Centres' for 'information'

'Discharging Centres'. Impulses sent out to modify MUSCULAR CONTRACTION.

Through these wide-spread connections the *EXTRAPYRAMIDAL SYSTEM* exercises a STEADYING influence on MUSCLE TONE.

Concerned in Control of MUSCLE TONE and CO-ORDINATION, and in POSTURAL REFLEXES.

Some fibres carry impulses to <u>restrain</u> Lower Motor Neurone activity – others to <u>facilitate</u> it.

Interplay of impulses
↓
SMOOTH CO-ORDINATED MUSCLE MOVEMENT

Little is known about the methods for regulating this interplay of impulses converging on the MOTOR UNIT during a voluntary action.

LOCOMOTOR SYSTEM

BONES and MUSCLES ——— concerned with MOVEMENT of the body.

SKELETON——RIGID FRAMEWORK gives SHAPE and SUPPORT to body.
is JOINTED to permit MOVEMENT

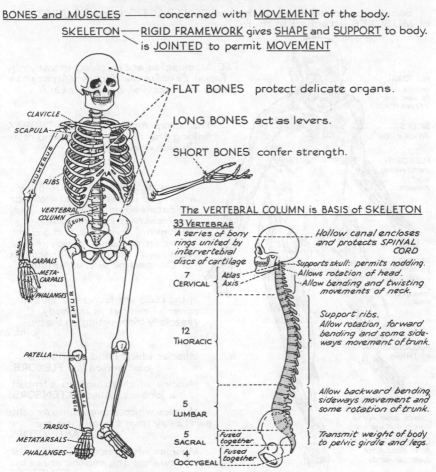

FLAT BONES protect delicate organs.

LONG BONES act as levers.

SHORT BONES confer strength.

CLAVICLE
SCAPULA
HUMERUS
RIBS
VERTEBRAL COLUMN
ILEUM
ULNA
RADIUS
CARPALS
META-CARPALS
PHALANGES
FEMUR
PATELLA
FIBULA
TIBIA
TARSUS
METATARSALS
PHALANGES

The VERTEBRAL COLUMN is BASIS of SKELETON

33 VERTEBRAE

A series of bony rings united by intervertebral discs of cartilage

Hollow canal encloses and protects SPINAL CORD

7 CERVICAL	Atlas	Supports skull: permits nodding.
	Axis	Allows rotation of head.
		Allow bending and twisting movements of neck.
12 THORACIC		Support ribs. Allow rotation, forward bending and some sideways movement of trunk.
5 LUMBAR		Allow backward bending sideways movement and some rotation of trunk.
5 SACRAL	Fused together	Transmit weight of body to pelvic girdle and legs.
4 COCCYGEAL	Fused together	

All bones give attachment to muscles.

SKELETAL MUSCLES

Bones are moved at joints by the CONTRACTION and RELAXATION of MUSCLES attached to them.

PECTORAL
brings arm to side and across chest

BICEPS
bends elbow

FLEXORS
bend wrist and fingers

RECTUS FEMORIS
bends hip-joint and straightens knee

ADDUCTORS of THIGH

SARTORIUS
bends knee and hip-joints and turns thigh outwards

EXTENSORS
turn foot and toes upwards

FACIAL muscles are involved in varying facial *EXPRESSION; SPEECH; MASTICATION* [Some muscles link bone to skin.]

The deep muscles of the THORAX, linking the ribs, contract and relax in *RESPIRATION*.

The muscles of the ABDOMEN are arranged in sheets and *PROTECT* delicate abdominal organs. They also contract to compress abdominal contents and aid in *MICTURITION, DEFAECATION, VOMITING* (and in the processes of *CHILDBIRTH* in the female).

In the LEGS are found the most powerful muscles of the body – especially those acting on the hip-joint.

Muscles which bend a limb at a joint are called **FLEXORS**.

Muscles which straighten a limb at a joint are called **EXTENSORS**.

Muscles which move a limb (or other part) away from the midline are called **ABDUCTORS**.

Muscles which move a limb (or other part) towards the midline are called **ADDUCTORS**.

146

SKELETAL MUSCLES

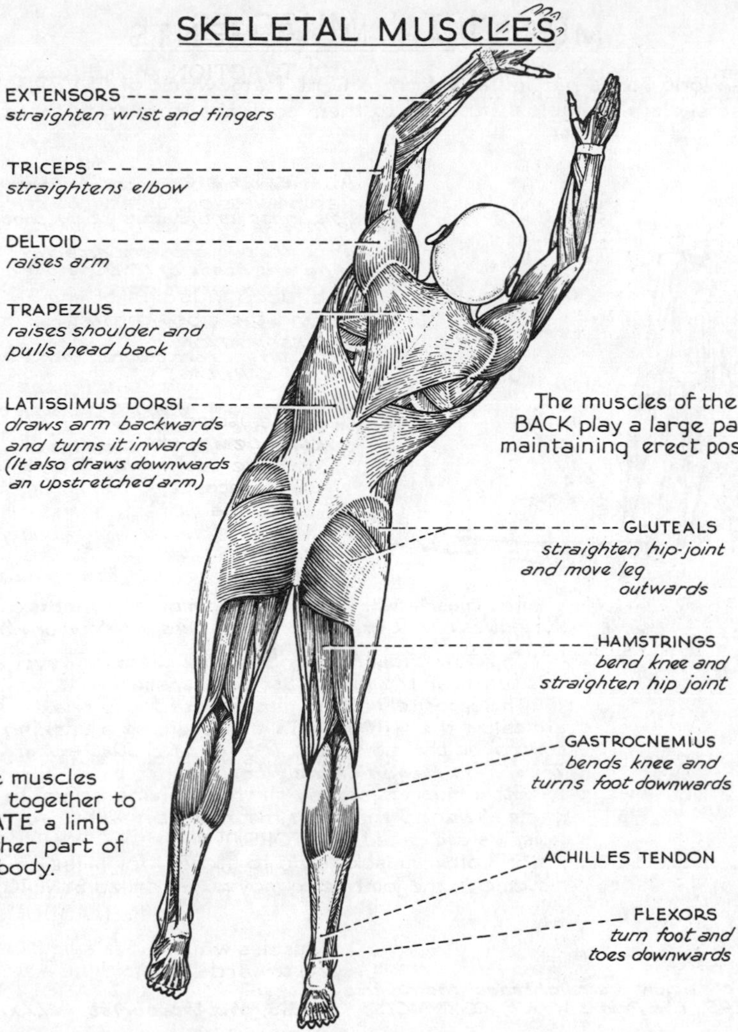

EXTENSORS -------------------
straighten wrist and fingers

TRICEPS -------------------
straightens elbow

DELTOID -------------------
raises arm

TRAPEZIUS -------------------
*raises shoulder and
pulls head back*

LATISSIMUS DORSI --------
*draws arm backwards
and turns it inwards
(It also draws downwards
an upstretched arm)*

The muscles of the
BACK play a large part in
maintaining erect posture

--------- GLUTEALS
*straighten hip-joint
and move leg
outwards*

---------- HAMSTRINGS
*bend knee and
straighten hip joint*

GASTROCNEMIUS
*bends knee and
turns foot downwards*

ACHILLES TENDON

FLEXORS
*turn foot and
toes downwards*

Some muscles
work together to
ROTATE a limb
or other part of
the body.

147

MUSCULAR MOVEMENTS

The long bones particularly form a light framework of LEVERS.
The skeletal muscles attached to them contract to operate these levers.

When a muscle contracts it shortens.

This brings its two ends closer together.

Since the two ends are attached to different bones by TENDONS one or other of the bones must move.

Two bones meet or articulate at a JOINT.

Joint surfaces are covered with a layer of smooth CARTILAGE.

To avoid friction when the two surfaces move on one another a SYNOVIAL MEMBRANE secretes a LUBRICATING FLUID.

BICEPS

TRICEPS

FLEXION

EXTENSION

The muscle which contracts to move the joint is called the PRIME MOVER or AGONIST *(the Biceps in flexion of elbow)*

To allow the movement to take place, however, other muscles near the joint must co-operate :-

The oppositely acting muscles gradually relax — these are called the ANTAGONISTS and exercise a "braking" control on the movement. *(e.g. the EXTENSORS — chiefly Triceps - in flexion of the elbow.)*

Other muscles steady the bone giving "origin" to the Prime Mover so that only the "insertion" will move— these muscles are called FIXATORS.

Still other muscles help to steady, for most efficient movement, the joint being moved — called SYNERGISTS.

When the elbow is straightened the reverse occurs:-
TRICEPS, the Prime Mover, CONTRACTS;   BICEPS, the Antagonist, RELAXES.

148

AUTONOMIC NERVOUS SYSTEM

The Autonomic Nervous System is concerned with maintaining a <u>Stable Internal Environment</u>. It governs functions which are normally carried out below the level of consciousness.

It has 2 separate parts – <u>PARASYMPATHETIC</u> and <u>SYMPATHETIC</u>. Both have their Highest Centres in the Brain.

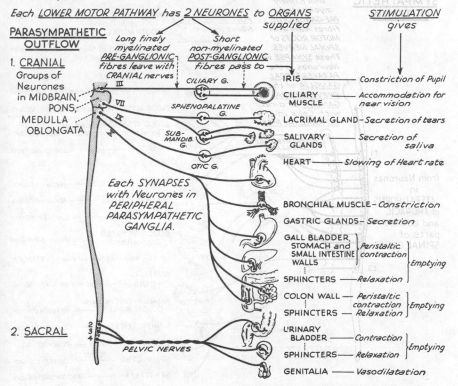

Each <u>LOWER MOTOR PATHWAY</u> has <u>2 NEURONES</u> to <u>ORGANS</u> | *<u>STIMULATION</u>*
supplied | *gives*

PARASYMPATHETIC OUTFLOW

Long finely myelinated <u>PRE-GANGLIONIC</u> fibres leave with CRANIAL nerves → *Short non-myelinated <u>POST-GANGLIONIC</u> fibres pass to*

1. <u>CRANIAL</u>
Groups of Neurones in MIDBRAIN, PONS, MEDULLA OBLONGATA

- III — CILIARY G. — IRIS — *Constriction of Pupil*
- VII — SPHENOPALATINE G. — CILIARY MUSCLE — *Accommodation for near vision*
- IX — — LACRIMAL GLAND – *Secretion of tears*
- SUBMANDIB. G. — SALIVARY GLANDS — *Secretion of saliva*
- OTIC G. — HEART — *Slowing of Heart rate*

Each SYNAPSES with Neurones in PERIPHERAL PARASYMPATHETIC GANGLIA.

- BRONCHIAL MUSCLE – *Constriction*
- GASTRIC GLANDS – *Secretion*
- GALL BLADDER, STOMACH and SMALL INTESTINE WALLS — *Peristaltic contraction* } *Emptying*
- SPHINCTERS — *Relaxation*
- COLON WALL — *Peristaltic contraction* } *Emptying*
- SPHINCTERS — *Relaxation*

2. <u>SACRAL</u>
2 3 4
PELVIC NERVES

- URINARY BLADDER — *Contraction* } *Emptying*
- SPHINCTERS — *Relaxation*
- GENITALIA — *Vasodilatation*

General stimulation of the Parasympathetic promotes vegetative functions of the body. Stimulation of the Parasympathetic and inhibition of Sympathetic have the same overall effects.

149

AUTONOMIC NERVOUS SYSTEM

The Parasympathetic and Sympathetic systems normally act in balanced reciprocal fashion. The activity of an organ at any one time is the result of the two opposing influences.

Each <u>LOWER MOTOR PATHWAY</u> has <u>2 NEURONES</u> to <u>ORGANS</u> supplied

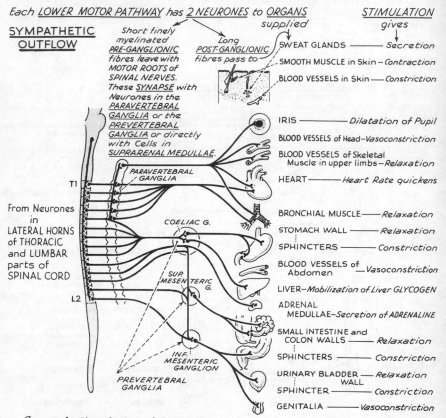

STIMULATION gives

SYMPATHETIC OUTFLOW

Short finely myelinated <u>PRE-GANGLIONIC</u> fibres leave with MOTOR ROOTS of SPINAL NERVES. These <u>SYNAPSE</u> with Neurones in the <u>PARAVERTEBRAL GANGLIA</u> or the <u>PREVERTEBRAL GANGLIA</u> or directly with Cells in <u>SUPRARENAL MEDULLAE.</u>

Long POST-GANGLIONIC fibres pass to

SWEAT GLANDS —— *Secretion*
SMOOTH MUSCLE in Skin — *Contraction*
BLOOD VESSELS in Skin —— *Constriction*

IRIS ———— *Dilatation of Pupil*
BLOOD VESSELS of Head– *Vasoconstriction*
BLOOD VESSELS of Skeletal Muscle in upper limbs– *Relaxation*
HEART ———— *Heart Rate quickens*

PARAVERTEBRAL GANGLIA

T1

From Neurones in LATERAL HORNS of THORACIC and LUMBAR parts of SPINAL CORD

COELIAC G.

SUP. MESENTERIC G.

L2

INF. MESENTERIC GANGLION

PREVERTEBRAL GANGLIA

BRONCHIAL MUSCLE —— *Relaxation*
STOMACH WALL —— *Relaxation*
SPHINCTERS ——— *Constriction*
BLOOD VESSELS of Abdomen —*Vasoconstriction*
LIVER– *Mobilization of Liver GLYCOGEN*
ADRENAL MEDULLAE– *Secretion of ADRENALINE*
SMALL INTESTINE and COLON WALLS —— *Relaxation*
SPHINCTERS ——— *Constriction*
URINARY BLADDER —— *Relaxation* WALL
SPHINCTER ——— *Constriction*
GENITALIA ——— *Vasoconstriction*

General stimulation of Sympathetic results in mobilization of resources to prepare body to meet emergencies. Stimulation of the Sympathetic and inhibition of the Parasympathetic have the same overall effects.

INDEX

Bold figures indicate main reference.

Breathing, mechanism of, 72
Broad ligament (uterine), 108
Bronchi, bronchioles, **68-70**, (149, 150)
Brunner's glands, 37
Bulbo-urethral glands, 103
Bundle of His, 52

C

Caecum, 27, 40
Calcitonin, 88
Calcium, 15, 19, 21, 32, **61**, (84), 88, **90, 91,** 93
Calorie, 19, 24, 25, 44
Capillaries, 47, 53, 57
Carbohydrate, 15, 16, 19, **20, 26,** 28, 32, 35, 37, 43, (87-99)
Carbon cycle, 17
Carbon dioxide (see Respiration)
Cardiac cycle, 50
Cardiac muscle, **11**, (47-52)
Cardiac sphincter, 27, 30, 31, **32**, (42)
Carotene, 22
Carotid body, sinus, 77
Cartilage, **10**, 69, 70, (137), 148
Cauda equina, 121
Caudate nucleus, 118, 124, 144
Cell, 1, 3, 4, 5
Centriole, centrosome, 3, 4
Cerebellum, (12), 118, 119, (139, 140), 144
Cerebral cortex, (12), **116-120,** 124, 127, 135, 136, 142-144
Cerebrospinal fluid, 67
Cervix, uterus, 108-111
Chemical senses, 126-128
Chemoreceptors, 77
Childbirth, 110-112
Chlorophyll, 16

Cholecystokinin/Pancreozymin, 35, 36, 37, 87
Choleretic action, 36
Cholesterol, 60, 89
Choline, 23
Chordae tendineae, 48
Chorionic villi, 109, 110
Choroid coat of eye, 129
Choroid plexuses, 67, **118**
Chromatin, 3, 4
Chromophils, chromophobes, of anterior pituitary, 95
Chromosomes, 3, 7
Chymotrypsin, 35
Cilia, 6, 69, (70), 108, 109
Ciliary body, ganglion, muscle, nerves, 120, 129, 131, 149
Circulation, 47
Claustrum, 118
Cleavage of zygote, 109
Clotting, 61
Cobalamine, 23
Coccygeal nerves, 121
Cochlea, (120), 137-139
Coeliac ganglion, 150
Cold, 46 (77), 141, (142)
Collagen fibres, 9
Colon, 27, **40, 41,** 42, (149, 150)
Colour blindness, 134
Conditioned reflexes, 29, **124**
Cones (retina), 132-134
Conjunctiva, 129, **130**
Connective tissues, **8-10,** (23)
Cornea, 129, **131**
Corpus albicans, 106
Corpus callosum, 119
Corpus luteum, 106, 107, 113
Corpus striatum, 118, 124, 144
Corticoids, adrenal, (82, 83), **92, 93,** 95
Corticotrophin, 92, 95, 97
Cranial nerves, 120
Creatinine, 60, 81, 84
Cretin, 89
Crypts of Lieberkuhn, 37

Cumulus oophorus, 106
Cuneate nucleus, 142
Cushing's syndrome, 93, 97
Cytoplasm, 1, 3

D

Decidua, 111
Defaecation, 40, 41
Dendrites, 12, 13, 122
Dense fibrous tissue, 9
Deoxycorticosterone, 92
Deoxyribonucleic acid (DNA), 3
Diabetes insipidus, mellitus, 98, 99
Diaphragm, (68), 71, 72, 76, 77
Diastole, 50, 51, 53-56
Diet, 19-26, (62)
Digestion, 27-39
Diiodotyrosine, 88
Diuresis, 82
Duodenum, 27, **35-39**, 42, 43, (62)
Dura mater, 67
Dwarfing, 89, 96

E

Ear, 137-139
Elastic fibres, **9**, 10, 53, 56, 69, 72
Embryo, 109, 110
Endocardium, 48
Endocrine glands, 87-100
Endolymph, 138, 139
Endolymphatic duct, 139
Endometrium, 108-111, 113
Endoplasmic reticulum, 3
Endothelium, (6), 48, 53
Energy sources, requirements, uses, 20, 24, 25, 44, (88, 89)
Enterogastrone, 33, 34
Enterokinase, 35, 37

152

Enzymes, 27, 28, 32, 33, 35, 37, 60, **62**
Eosinophils, 8, **90**, 95
Epididymis, 101, 102
Epithelia, 6, 7
Equilibrium, organs of, 126, 137, **139**
Ergastoplasm, 3
Erythrocytes (see Red blood corpuscle)
Erythropoietin, 62
Eustachian tube, 137
Excretion, (2), **78**
Exophthalmos, 89
External auditory meatus, 137
Exteroceptors, 126
Extrapyramidal system, 144
Extrinsic factor, 62
Extrinsic muscles of eye, (120), 130
Eyes, (120), 129-136, (149, 150)

F

Facial muscles, nerve (VII), 120
Factors V, VII, VIII, IX, X, 61
Faeces, 36, 40, 41, 44, 59
Fallopian tubes (uterine tubes), (6), **108, 109**, 114
Fat, 15, 16, **19, 20**, 26, 32, 35, 36, 37, 39, 43, 45, 46, (99)
Fatty acids, 19, 35, 36, 37, 39, 43
Female (see Reproductive system)
Fertilization, 109, 113
Fibrin(-ogen), 60, 61
Fibrous tissue, 9
'Fight or Flight' mechanism, 94
Filtration (see Kidney)
Fixators, 148
Foetus, (64), **110**
Folic acid, 23, 62

Follicle-stimulating hormone (FSH), 95, 113
Food, 19-26, (27-43)
Fovea centralis, 132
Frohlich's dwarf, 96
Fundus oculi, 132
Fundus of stomach, 32, 34

G

Gall bladder, 27, 36, (149)
Gastric (glands) juice, 32, 33
Gastrin, 33
Gastro-colic reflex, 41
Gastro-intestinal hormones, 33-37, 39, (87)
Gastro-intestinal tract, 27-42
Genes, 3, 4
Geniculate bodies, 135
Giantism, 97
Glands (see under specific sites)
Globulin, 60, 65
Globus pallidus, 118, 144
Glomerulus, kidney, 78-81
Glossopharyngeal nerve (IX), 120
Glottis, 69, 77
Glucagon, 99
Glucocorticoids, 92, 93
Glucose, 20, 32, 35, 37, 39, 43, 81, 92, 93, 99
Glycerides, 35-37, 39
Glycerol, 37, 39, 43
Glycogen, **20**, 43, 92, 94, 99, (108)
Goblet cells, **6**, 40, 69
Golgi body, 3
Golgi organ (in tendons), 140
Gonadotrophins, **95**, 104, 113-115
Gonadotrophin releasing factor, 104, 113, 114
Graafian follicle, **106**, 113, 114
Gracile nucleus, 142
Granular leucocytes, **8**, 62

Grey matter, 117, 118, 121
Growth, (19), 21-24, 25, 43, 87-100, 104, 114
Growth hormone, 95-97
Gustatory cells, 128

H

Haemoglobin, 15, 21, 62, (63), 75
Haemolysis, 63, 64
Haemolytic disease of newborn, 64
Haemopoiesis, 62, 65, 66
Haemopoietic factor, 62
Haemostasis, 61
Hair, 45, 46, 89, 93, 94, 97, 101, 104, 105, 114, 115, **141**
Hassall's corpuscles, 100
Haversian canal (in bone), 10
Hearing, 117, 120, 137, 138
Heart, 11, **47-58**, (89, 94, 97, 149, 150)
Heat, 44-46, (89), 141
Henle's loop (kidney), 79, 81
Heparin, 8
Hering-Breuer reflex, 77
Hirsutism, 93, 97
Hormones, 33-37, 39, 87-115
Hyaloplasm, 3
Hydrochloric acid, in gastric juice, 32, 33
Hydrocortisone, 92, 93
Hypermetropia, 131
Hypoglossal nerve (XII), 120
Hypothalamic-releasing factors, 106
Hypothalamus, 45, 46, 82, 83, 88, 92, 94, 95, 98, 111, 112, 113, 119, 144

I

Ileo-caecal valve, 27, 37, 38, 40, 41, 42
Ileum, 27, 37

Immunity, 8, (60), 65, **100**
Implantation, 109, 110
Incisura angularis, 34
Incus, centres for smell, 127
Incus, middle ear, 137
Inferior mesenteric ganglion, 150
Inositol, 23
Inspiration, **72-77**
Insula, 118
Insulin, 99
Intercalated discs, 11
Intercostal muscles, 71, 72
Internal capsule, 118, 142, 143
Interoceptors, 126
Interstitial cells, 102, 104, (114)
Intervertebral discs, 10, 145
Intestine, (6), 27, 34-43, 90, 91, (149), 150
Intrapleural intrathoracic, pressure, (58), 71, 72
Intrinsic factor, 32, 62
Involuntary muscle, 11
Iodine, 21, 26, 88
Iris, 94, **129**, (149, 150)
Iron, 15, 21, 26, 32, (43), 60, **62**
Islets of Langerhans, 35, **99**

J

Jejunum, 27, 37, 38, 39, 42, 43
Joints, 140, 145, **148**

K

Ketone bodies, 99
Kidney, **78-83**, 90-93, (97), 98, 99
Krause bulbs, in skin, 141
Kupffer cells (liver), 36

L

Labyrinth, 139
Lacrimal ducts, glands, sac, 130
Lactase, lactose, 37
Lactation, 25, 95, **112**
Lacteals, 39, 43
Large intestine, 27, **40-43**, (149, 150)
Larynx, (31), 68, **69**, 120
Lemnisci, 142
Lens of eye, 129, 131
Leucocytes, leucopoiesis, 8, 60, 62, 65, 66
Leydig, cells of, 102
Ligaments, histology of, 9
Lipase, 32, 35-37
Liquor folliculi, 106
Liver, (20), (27), **36**, 43, (44), 62, 92-94, 99, 150
Locomotor system, 14, 145-148
Loop of Henle, 79, 81
Loose fibrous tissue, 9
Lorain dwarf, 96
Lumbar nerves, 121
Lungs, 44, 47, 59, **68-77**, (149, 150)
Luteinizing hormone (LH), 95, 104, 113-115
Lymph, nodes, 8, 43, **65**
Lymphatics, 39, 43, 65
Lymphocytes, lymphoid tissue, 8, 37, 40, 65, 66, 100
Lysosomes, 3

M

Macrophage cells, (8), 65, (66)
Macula lutea, 132
Male, (92), 101-104
Malleus, 137, (138)
Maltase, maltose, (28), 35, 37
Mammary glands, (95, 98, 105), 112, (114), 115

Mastication, 28
Medulla oblongata, 29, 31, 33, 42, 76, 77, **118-121**, **142**, **143**, **144**, **149**, (150)
Meiosis, 4, 102, 106
Meissner corpuscles, in skin, 13, 141
Meissner's nerve plexus, in gut, 39
Membrana granulosa, 106
Meningeal membrane, 67
Menopause, 115
Menstrual cycle, 106, 108, 113, 114, 115
Mesentery, of gut, 39, 40
Metabolism, 23, 24, 25, 43, 44, (59), 81, 87-99
Metaphase, 4
Microvilli, 37
Micturition, 86
Midbrain, 118-120, 143, 144, 149
Milk, 20-23, 26, 32, 35, (112)
Mineralocorticoids, 92, 93, 97
Minerals, 15, 19, 21, 26, 32, 39, (40), 43, 57, 59, (81), 82, 83, 88, 90-93
Mitochondria, 3
Mitosis, 4, (102)
Mitral valve, 48, 51
Monochromatic vision, 133
Monocyte, 8
Monoiodotyrosine, 88
Motor end-plates, 12, 143
Motor nerve cells, 12, 117, (121), (143)
Mouth, 28
Mucous glands, 28, 30, 69
Multipolar neurones, 12
Muscles, 11, 12, 45, 46, 94, 140, 143, 144, 146-148 (see under Smooth, Skeletal and Cardiac muscle)
Muscle spindle, (123), 140
Muscularis mucosae, 39
Myelin, 12

154